CONTROL THEORY METHODS IN ECONOMICS

CONTROL THEORY METHODS IN ECONOMICS

by

JATI K. SENGUPTA

Professor of Economics and Operations Research,
University of California, Santa Barbara

and

PHILLIP FANCHON

Associate Professor of Economics,
California Polytechnic, San Luis Obispo

SPRINGER SCIENCE+BUSINESS MEDIA, LLC

ISBN 978-0-7923-9845-5 ISBN 978-1-4615-6285-6 (eBook)
DOI 10.1007/978-1-4615-6285-6

Library of Congress Cataloging-in-Publication Data

A C.I.P. Catalogue record for this book is available from the Library of Congress.

Printed on acid-free paper

To
Krishna, Rimi and Raja

and
Elaine, Ira and Alosha.

"Not everything that counts
can be counted
Not everything that can be
counted counts."

Albert Einstein.

CONTENTS

5 ECONOMIC IMPLICATIONS OF STOCHASTIC CONTROL

6 VARIABLE STRUCTURE SYSTEMS

7 RISK SENSITIVITY, ADJUSTMENT OF CONTROL AND ESTIMATION

Preface

Control theory methods in economics have historically developed over three phases. The first involved basically the feedback control rules in a deterministic framework which were applied in macrodynamic models for analyzing stabilization policies. The second phase raised the issues of various types of inconsistencies in deterministic optimal control models due to changing information and other aspects of stochasticity. Rational expectations models have been extensively used in this plan to resolve some of the inconsistency problems. The third phase has recently focused on the various aspects of adaptive control, where stochasticity and information-adaptivity are introduced in diverse ways e.g., risk adjustment and risk sensitivity of optimal control, recursive updating rules via Kalman filtering and weighted recursive least squares and variable structure control methods in nonlinear framework. Problems of efficient econometric estimation of optimal control models have now acquired significant importance.

This monograph provides an integrated view of control theory methods, synthesizing the three phases from feedback control to stochastic control and from stochastic control to adaptive control. Aspects of econometric estimation are strongly emphasized here, since these are very important in empirical applications in economics.

Besides containing the recent developments in applied control theory, the monograph discusses several new results from our joint research over the past decade. These include among others the following: (1) variable structure control, (2) adaptive and risk-sensitive control, (3) estimation of Euler equations in Optimal control models, including the transversality condition, (4) consistency problems in differential game models, and (5) some of the recent economic applications e.g., limit pricing models, portfolio efficiency frontier and dynamic adjustment models arising in optimal growth theory.

Jati Sengupta
Phillip Fanchon

CHAPTER 1: **Introduction to Control Theory**

1.1 Transition from the calculus of variations to optimal control

The calculus of variations was born from the problem of finding the curve joining two points such that a body sliding along it without friction would move from one end to the other in minimum time. The so-called "brachistochrone" problem, proposed over 250 years ago, is part of the general class of problems of finding the extreme values of a functional of the form:

$$J = \int_a^b f(x, x', t)dt$$

where f is a continuous function of x, x' (= dx/dt) and t. An elegant method of solution is to add a disturbance ("variation") of the form $\varepsilon\eta(t)$ to the variable x^* following the optimal path. If x^* is indeed the optimal solution, then it must hold that $dJ(\varepsilon)/d\varepsilon = 0$ at $\varepsilon = 0$. With the continuity assumption for f, differentiation through the integral sign yields the so called Euler equations (also known as the Euler-Lagrange equations)

$$\frac{\partial f}{\partial x} - \frac{d}{dt}\left(\frac{\partial f}{\partial x'}\right) = 0,$$

which identify the first order conditions for the extremum of the functional. The Euler equations, and the associated solution derived, depend upon two important assumptions; (i) the integrand being maximized (or minimized) is a continuous function of x, x', and t, and (ii) the integral is finite. These two conditions are essential for the validity of the solution as originally proposed. Without finite bounds, the differential equation which is the by-product of the Euler equations cannot be solved. Two complications are common in economics and finance problems The first one is that the initial conditions might be very difficult to measure (e.g., some of the data might not be readily available, or measuring the initial conditions might take a long time). The second one arises when the planning horizon is infinite. Since the integral must remain bounded, some restrictions must be imposed as time becomes infinite. Hence some "terminal conditions" are needed to guarantee that the integral does indeed remain bounded.

The limitations of the calculus of variations are apparent; The two main weaknesses are that (1) there is no explicit control variable u in the original solution that would allow for the direct intervention of a controller, and (2) the assumption of continuity and boundedness must hold. Two methods have been developed to deal with the first weakness, and both introduce a control variable u, and a state equation which describes the dynamics of the system. The first method is to use the state equation to express the control variable in terms of the state variable $u = g(x)$, and then substitute for the control variable in the integrand. The problem is at this point reduced to the original problem of the calculus of variations and the state trajectory can then be derived by solving the Euler equations. The second method, used when it is not possible to express the control in terms of the state variable, is to introduce the state equation as a constraint within the maximand. The maximand is augmented by the conditions of the state equation, multiplied by a "dynamic" Lagrange multiplier λ, called a "costate variable. Thus, the problem is reduced to a special type of calculus of variations problem, where there are now three variables (x, u, λ, and t), instead of the original two (x and t). Nevertheless, we can still use the calculus of variations approach, and use the Euler equations three times; once for the state variable, once for the costate variable, and once for the control variable. The resulting set of differential equations must then be solved jointly to identify the solution. This approach is the one followed by the vast majority of early applications of "control theory" in economics. These two methods reduce the original problem to a variational calculus problem solved using the Euler equations. The second weakness cannot be overcome; The conditions that must hold for the original problem must still hold for the transformed problem, i.e. the integral must be bounded and the maximand must be continuous in x, x', and t. The continuity condition is particularly restrictive since many possible control actions are not continuous. For example, to move a car from one point to another in minimum time requires a discontinuous control action; maximum acceleration followed by maximum deceleration (braking), clearly a discontinuous action.

Control theory made a giant leap forward with the formulation of the "Maximum Principle", which spells out first order conditions that must be met by the modified maximand (the Hamiltonian) to identify an extremum of the integral. The importance of the contribution is not obvious for the class of problems where the control action is unbounded and continuous because the conditions of the maximum principle are identical to the set of differential equations obtained in the second approach described above. The difference becomes obvious for problems with discontinuous control, like moving a car from one point to another, which can be solved with the conditions of the maximum principle, and cannot be solved with the variational calculus approach.

The bulk of theoretical work in optimal control has been done as an extension of the maximum principle, which is set in the continuous time framework (i.e. the control action can be discontinuous, but the evolution of all the variables is continuous in time). As a result, most of the early applications were also set in continuous time. In physics and engineering, it was relatively easy to approximate the conditions of continuous time by monitoring the system performance with either analog devices or high frequency sampling. Economists followed the same path, assuming that policy makers have instant and continuous knowledge of the state of the system they want to control. These convenient simplifying assumptions are dying off very slowly, and from the initial applications of control methods to economics e.g., Arrow and Kurz (1971), to the recent past e.g., Seierstad and Sydsaeter (1987), Chiang (1992), and Turnovsky (1995), the majority of the research and textbooks has cast in the continuous time frame.

Since economic and financial data are collected at discrete points in time, the tools of differential calculus are of little help for the decision maker using discrete data. In particular, the assumptions of continuity made in the calculus of variations and in the original formulation of the maximum principle cannot be extended to applied economics and econometrics. Instead, one must use lesser known results in difference equations. A major step towards modeling reality was made with the derivation of the discrete form of the maximum principle. The gap between theoretical control and applied economics was then nearly bridged, and numerous macroeconomic models using real data soon appeared e.g., Chow (1975) and Kendrick (1979, 1981). Many difficult issues remain, roughly lumped in two groups; 1/ estimation and specification problems (order selection, modeling and information errors, nonlinearities and unmodelled dynamics, control by several controllers, estimation versus control, time inconsistencies, forward versus backward looking models, robust versus optimal control), and 2/ asymptotic model behavior (transversality conditions specification, chaotic behavior and limit cycles). We certainly cannot cover all issues, especially since many of these issues are currently the object of very active research.

Building upon a modest knowledge of calculus and linear algebra, this book follows the evolution of control theory from the calculus of variations to some current issues. Starting with the continuous case, we present the essential results without dwelling on proofs since numerous books are available with elaborate proofs. This shortcoming is a necessary evil due to the scope of the book, which is both to provide the necessary elements needed to understand basic control papers, and also to present through applied work some recent advances and outstanding problems in the field. Chapters 2 and 3 cover the essential knowledge for solving continuous and discrete deterministic control problems. The rest of the book is focused on applied stochastic models and on some of the issues mentioned above. Throughout the book, we offer applied examples on how to resolve those issues.

1.2 Transition from theory to practice

In recent years, applied control theorists were greatly helped by the stunning leap in the processing power of small computers. The quest for an accurate method of solution for large problems now seems more feasible as the cost of computing decreases, and iterative methods are gaining acceptance as a way to identify new distributions, or to solve complex numerical problems. The advantage is that it allows for the identification of the solution of difficult problems when there is no analytical way to solve them. The growing number of simulations in published research is a testimony to this trend. We use this powerful method in chapter 2 to identify the optimal policy for a firm training its own labor. The theoretical results of the proposed model are confirmed by the results of several simulations, which also validate the practice of many firms who provide cyclical training to their employees. The simulation results point directly to one of the nagging problems associated with using an infinite horizon; How to guarantee that the controlled system remains stable as time becomes infinite? The common practice among applied control theorists is to identify the roots of a linearized version of the model used, and to ignore the roots that yield an explosive behavior in the design of the optimal control. Such a practice amounts to throwing away some information about the system dynamics, and to identify a control law that is optimal in a system of reduced. The simulation clearly shows that such a practice might lead to erroneous results, especially in the case where the continuity assumptions of the Euler equations do not hold.

Stochasticity is introduced in chapter 4, where the first economic application defines the main tool used in applied stochastic control; the linear quadratic gaussian (LQG) control model. The first application also points to the dilemma faced by the controller; how to divide time and resources between estimation and control. Another dilemma faced by the controller is the choice of a model. The second application illustrates the dramatic jump in complexity when the interaction between several controllers is incorporated in the model. The differential game formulations of competition between two countries or two firms can be solved explicitly under various economic scenarios only when considering two periods. Despite some potential estimation difficulties, the maximum likelihood estimates of the parameters yields a feedback rule that is asymptotically optimal (i.e. the estimated rule converges to the optimal rule as the number of observations increases). The third example illustrates in particular how the LQG formulation could yield a non optimal decision rule if the third and fourth moments of the distribution of the error term are ignored. The next section illustrates through numerous examples how an adaptive updating scheme (the Kalman filter) can be implemented to improve model performance. The various examples given show in detail how the controller can incorporate risk aversion into the objective function, and how the resulting control law becomes more cautious than in the purely deterministic models. Using a dynamic portfolio model, we demonstrate how using the Box-Cox transformation can reduce the original model to one where the stochastic properties of the random terms satisfy the normality assumptions of the LQG formulation.

The following chapter points to numerous problems introduced with the stochastic formulation: Because the parameters and the functional form of the model used are uncertain, the controller must constantly re-estimate the model, and formulate a policy that reflects the latest update. When the separation of estimation and control fails, (i.e. the control and estimation processes are jointly dependent), both the target and the control must be continuously updated through the method of adaptive control introduced in the chapter. In a large part, the need for continuous updating is the result of a conflict of objectives; Most often, the controller's decision is based not only upon past experience, but also upon expectations about the future. Hence most models should contain both a forward and a backward looking component, i.e. the models must also describe how the controller's expectations are formed and updated, a daunting task which is mostly omitted. Such an omission can be the main source of joint dependence between estimation and control, and an important source of bias.

Another difficult issue addressed in chapter 5 is how to address the conflicting objectives of maximization of a unique functional over two time horizons. For example, a firm might have two objectives; maximize profit in the short run, and maximize profit in the long run. The chapter introduces the notion of dynamic efficiency and illustrates how the time horizon conflict can be mitigated by recasting the problem as a two stage optimization problem. Finally, we offer an applied example of the two stage process using a growth model for Japan, cast in the familiar LQG framework as a tracking problem with a risk aversion component included in the objective function. The equations for the system dynamics are estimated in two ways; one is with the forward looking approach, where future values of the state vector are used to estimate the parameters of the current dynamic equations, and the other is with the backward looking approach where only past values are used to estimate the dynamic equations. The difference in estimates is then used to gauge the relative weight of the two objectives (short term versus long term) in the objective function used. The model proposed confirms Krugman's argument (1991) that for Japan, future expectations play a stronger role than past history in explaining their growth rate in capital investment and GDP. Econometric evidence also supports the theory that the adjustment lags and expectation formation are quite different for the various factors of production.

The difficulties encountered in estimation and control of a complex model can often be overcome with sophisticated techniques like the ones presented in chapters 4 and 5. However, to guarantee a control law that is optimal in some sense, one must use complex econometric techniques, which can be time consuming. In large decentralized systems, the quest for optimality might become quite elusive when agents are asked to implement control laws that are constantly changing. An alternate approach is to abandon the goal of optimality, and instead use a control law that will make the system behave in a desired manner, despite modeling inaccuracies or changing system parameters. The trade off is simply to decrease the complexity of the control law at the expense of optimality. The variable structure control approach presented in chapter 6 is one of the many possible avenues to achieve that goal. The technique assumes that at some point the controller has identified an optimal control law, and that instead of continuously updating the law, the controller decides to implement a sub-optimal control law that will maintain the state trajectory on or close to the path the state vector was expected to follow at the time the estimation was made (i.e. we solve a tracking problem). Hence the main attribute of such a sub-optimal control law is that it must be robust to misspecification or random errors while maintaining the original system dynamics.

We illustrate the use of variable structure control design using a simple second order adaptive expectation model, and compare the performance of the VSC law with that of the optimal control law. The simulation demonstrates that poor estimation of the system parameters can easily make the system unstable under an optimal control law while the VSC law will yields a "well behaved" system. Hence the claim that under certain conditions, robustness is preferable over optimality.

Finally, we introduce in chapter 7 some recent developments in control and estimation. The first part presents some adjustment cost models which have been proposed recently to provide an alternative to the mean variance efficiency model e.g., the adjustment cost approach developed by Green (1988) and others, who derive and empirically estimate of the asset price expectation equations by inverting a set of mean-variance-based asset demand functions, and the ARCH & GARCH formulations. An application to the returns of mutual funds in various markets over a 10 year period illustrates the estimation techniques and compared the performance of the models, which are adaptive in two ways; One is through updating the parameter estimates of the time varying system, and the other is through the inclusion of risk in the objective function, which influences the optimal decision rule through an estimate of the error covariance matrix.

The second part centers on estimation; After a brief exposition of the weighted recursive least squares algorithm, and of two methods of joint estimation and control (one by Ljung et.al. (1974), and the other by Chow(1975)), we apply the methods to a second order time varying model. The final part of the chapter returns to one of the initial problems faced by modellers considering an infinite horizon; estimating the transversality conditions. The conditions are necessary to fully identify the solution of an optimal control problem under the assumption that a steady state has been reached at the terminal time, assumption which itself depends upon the constraints imposed on the system and on the control. Note that the transversality conditions are not always needed, and that there are many instances where the system dynamics lead to a steady cycle instead of a steady state. Several examples are available in biology, chemistry, engineering, and are generally called survival models. Most the models without a steady state have two things in common; they are non linear, and involve a system of dimension greater or equal to two. Several cyclical models have been proposed in economics, most of them related to business cycles (the reader can consult the work of Richard Goodwin for an abundant source of examples and references). Clearly, any LQG model is an unlikely candidate for having a steady cycle as a steady state solution. yet, it does not imply that it has a steady state. For example consider a dynamic rational expectations model with dynamics: $y_t = ay_{t-1} + bu_t + dy_{t+1} + \varepsilon_t$.
The effect of the coefficient d is very critical in the long run as the terminal time becomes infinitely large. If it makes the system unstable in the sense that at least one

characteristic root exceeds one in absolute value, then imposing a fixed terminal value may not be consistent with the optimal trajectory. On the other hand, if the coefficient d preserves the stability of the system, then a fixed terminal condition can be added without much difficulty. To illustrate some of the basic issues posed by the terminal conditions, we consider now two economic examples, one from the dynamic adjustment model discussed in chapter 5, and the second from the neoclassical theory of optimal growth, and propose several methods of estimation of the steady state.

The book ends on a hopeful note; As new tools are being developed, the old ones should not be ignored (for example, the tools of "classical" control theory, based on the Laplace transformation which maps the initial problem in the "state space" to an equivalent problem in the "frequency domain", where the problem is solved). Classical control theory is still the main tool used in system design in engineering, its prevalence due to the fact that the controller can easily design a system with specific behavior. Until the very recent past, the transfer function approach was thought to be of little use to economists since it is generally used for single input single output systems. However, recent advances in the frequency domain approach have stimulated interest among systems engineers who have adapted the technique to multivariate systems. In particular, optimal control design based on the so-called H_∞ norm used in the frequency domain yields a control rule that is both optimal in some sense, and also robust to system and misspecification errors. The mathematical background needed for H_∞ control design is beyond the scope of this book, and readers interested in the topic could consult some of the earlier works by Doyle, or Maciejowski (1989). Other fields of mathematics, like bifurcation theory, chaotic dynamics and complexity theory, entropy and information theory and even distributed parameter systems (where partial differential equation systems are used instead of the ordinary differential equation used to model the economic dynamics) are slowly being used in economic modeling, and an increasing number of nonlinear control models are being applied in modeling economic dynamics (for example, Brock, Hsieh and LeBaron (1991)). Invariably these new concepts would require developing efficient methods for econometric estimation, particularly for nonlinear systems. Simulation and experimental analysis of structural data appear to be very useful in this framework.

References

Arrow, K. and Kurz, M, (1971), *Public investment, the rate of return, and Optimal Fiscal Policy.* Baltimore: John Hopkins Press.

Brock, W., Hsieh, D., LeBaron, B., (1991), *Nonlinear Dynamics, Chaos, and Instability: Statistical Theory and Economic Evidence.* The MIT Press, Cambridge, Massachusetts.

Chiang, A., (1992), *Elements of Dynamic Optimization*, McGraw Hill.

Chow, G. (1975), *Analysis and Control of Dynamic Systems*, New York: John Wiley.

Green, C. (1988), "Adjustment Costs and Mean Variance Eficiency in U.K. Financial Markets." *Economic modelling in OECD countries*, London: Chapman and Hall.

Kendrick, D., (1979), "Adaptive Control of Macroeconomic Models with Measurement Error", in *Optimal Control for Economic Models*, Holly, Rüstem and Zarrop editors, St. Martin's Press, pages 204-227.

Kendrick, D., (1981), *Stochastic control for economic models*, McGraw Hill.

Ljung, L., and Soderstrom, T. (1974) "Identification of Linear Multivariable Systems Operating Under Linear Feedback Control." *IEEE Transactions on Automaic Control*, AC-19, pages 836-840

Maciejowski, J., (1989), *Multivariable Feedback Design*, Addison & Westley.

Seierstad, A. And Sydsaeter, K., (1987), *Optimal Control Theory with Economic Applications*, North Holland.

Turnovsky, S. (1979), "Alternative Passive Monetary Policies in an Inflationary Economy." *Journal of Macroeconomics*, Vol. 1, No. 1, pages 33-63.

CHAPTER 2: **Continuous time models**

2.0 Overview of control problems

In its most general form, the optimal control problem is to find a control vector $u(t)$ that will give an extremal value to a performance index J (called the *objective function*), when the state of the system being controlled is described by an n-dimensional state vector $x = (x_1, x_2, \ldots x_n)$ whose components evolve according to the *state equations*.

A multitude of possible optimal control problems are possible; For example, the controller can use a finite or infinite planning horizon, can face some restrictions from what are acceptable values for the state and control vectors, or can have an imperfect knowledge of the system being controlled[1]. Only a few of the possible formulations have been solved explicitly, most of them for the special case where the state equation is linear, and the Hamiltonian function associated with the problem is "nicely behaved" (i.e. satisfies some continuity, differentiability, and concavity assumptions). The more difficult problems are often solved by linearization, or by removing some of their troublesome characteristics.

The difficulties come from three areas, all linked to continuity and differentiability. The first one is that because most modellers of economic systems are very familiar with the calculus of variations (which uses some very restrictive continuity assumptions about the performance index used and about the state equation), they tend to view all optimal control problems as variational problems, and make the usual continuity assumptions. The second difficulty is that the solutions of many nonlinear differential equations are still unknown. The third difficulty is that the "maximum principle", the central pillar of modern optimal control, only spells out the conditions necessary for the existence of a solution, but does not provide a method to identify that solution. Many questions are still unresolved, and progress has been limited to cases where the objective function is continuous in the control, and where the Hamiltonian is piecewise differentiable. The conditions that must prevail for a steady state in infinite horizon problems is still uncertain, especially when the Hamiltonian function is piecewise continuous, but nowhere differentiable. The fourth difficult area is that of estimation: Social scientists do not have the luxury of being able to conduct experiments with known initial conditions, for a system ruled by physical rules. Identification of the relevant variables, of the functional form of the model, of the number of leads or lags to be included are still open questions in economics, as exemplified by the current lively debates on the formation of expectations and on the estimation of time series.

---

1 An extensive compilation of continuous time economic models can be found in Turnovsky (1995).

2.1 Observability and Controllability

The task of a controller (i.e. a policy maker or a manager) is to use a set of control actions to make a system behave in a desired manner. For example, a pilot can use five controls (ailerons, rudder, elevators, trust and flaps) to follow a desired flight path defined by 4 state variables (longitude, latitude, altitude and speed). Two elements are essential for the successful completion of the flight; First, the pilot must know his position and speed, and second, the pilot must have the means to affect all four state variables. Like a pilot controlling an aircraft, economists need to have at their disposal the relevant information and enough instruments to implement monetary or fiscal policy. In the terminology of control theorists, the system must be *observable* and *state controllable*. These two concepts are at the heart of the control problem since an optimal control vector cannot be defined if the system is not observable and controllable.

2.1.1 Observability

To steer a set of state variables to a specified target, or along a specified path, the controller must know what the components of the state vector are. When the state vector is not known completely, the controller must obtain estimates of its missing components. Information about the value of those components can be gathered through proxies called *observers* (or *filters* for the stochastic case). On an airliner, the pilot has telemetric instruments on board giving directly his or her position and speed. The instruments are the observers providing complete and almost perfect information. For the pilot of a small airplane, the observer is the set of readings from instruments giving an estimate of atmospheric pressure, air speed, temperature, and so on. Together, all the pieces of information allow the pilot of the small airplane to determine indirectly his location and speed. In both situations, whether flying an airliner or a small airplane, the position and speed of the airplane can be inferred from the available information, and the system is observable. Similarly, firms usually do not know precisely what is the demand for their product. To reach their objective (for example maximize profits or market share, or minimize cost), firms use a proxy for demand, identified with statistical tools (for example using regression analysis or sampling methods). The concept of observability can now be defined formally:

Definition: A system is called *observable* if the information set available over a finite period of time is sufficient to determine the value of all the components of the state vector at each point in time.

If the information provided is insufficient to determine the value of the components of the state vector, the controller must then divide his efforts between estimation of the missing components, and control of the system. Most of chapters 2 and 3 assume that the observers provide enough information to determine completely and exactly all the components of the state vector. The study of systems with incomplete and imperfect information is covered in chapters 4 to 7.

It is usually quite difficult to verify that a non linear system is observable. The task is much easier for a linear time invariant system of the form:

$$dx/dt = A\,x(t) + B\,u(t)\,, \qquad y(t) = C\,x(t)\,, \tag{2.1.1}$$

where: $x(t) \in X^n$ is a vector of state variables defined on a subset of $\Re^n$,
$u(t) \in U^m$ is a vector of control variables, defined on a subset of $\Re^m$,
$y(t) \in Y^r$ is an information vector (also called output vector) defined on a subset of $\Re^r$.

If the controller has complete and perfect information, the matrix C is the identity matrix. In practice it is more convenient to check for observability by computing the (nrxn) *observability matrix* of the system, defined as:

$$R = \begin{bmatrix} C \\ CA \\ CA^2 \\ \vdots \\ CA^{n-1} \end{bmatrix}, \tag{2.1.2}$$

and use the following theorem.

Theorem: The system (2.1.1) is observable if and only if its observability matrix has rank n.

Example 1:

Determine if the system $\begin{pmatrix} \dot{x}_1 \\ \dot{x}_2 \end{pmatrix} = \begin{bmatrix} 2 & 1 \\ 6 & 7 \end{bmatrix}\begin{pmatrix} x_1 \\ x_2 \end{pmatrix} + \begin{pmatrix} 3 \\ 1 \end{pmatrix} u$, $y = \begin{pmatrix} 6 & -1 \end{pmatrix}\begin{pmatrix} x_1 \\ x_2 \end{pmatrix}$ is observable.

The observability matrix is: $R = \begin{bmatrix} 6 & -1 \\ 6 & -1 \end{bmatrix}$. Since rank[$R$] = 1 < n , the system is unobservable.

Example 2:

Determine if the system $\begin{pmatrix}\dot{x}_1\\ \dot{x}_2\end{pmatrix}=\begin{bmatrix}2 & 1\\ 3 & 2\end{bmatrix}\begin{pmatrix}x_1\\ x_2\end{pmatrix}+\begin{pmatrix}3\\ 1\end{pmatrix}u$, $y=\begin{pmatrix}6 & -1\end{pmatrix}\begin{pmatrix}x_1\\ x_2\end{pmatrix}$ is observable.

The observability matrix is: $R=\begin{bmatrix}6 & -1\\ 9 & 4\end{bmatrix}$. Since rank[$R$] = 2 = n , the system is observable.

2.1.2 Controllability[2].

To reach a destination, an airline pilot must not only know where the airplane is, but also must have the means to fly the airplane. He or she can use the ailerons, rudder, elevators and fuel to control the position and speed of the aircraft. Even though it is possible to reach the destination without rudder, it is impossible to reach it without fuel or elevators. Policy makers trying to control inflation and unemployment, not only need to know price indices and the unemployment rates, but they must also have the tools of fiscal and monetary policy to affect the economy. In the language of control theorists, to be able to reach his objective, the controller must not only know or have enough information to estimate the current value of the state vector (i.e. the system must be observable), but also must have at his or her disposal enough instruments to make the system behave in a desired manner, starting from any possible initial state. When such conditions exist, the system is called *state controllable*[3]. The concept of controllability can now be defined formally:

Definition: A system is *state controllable at time* t_o if there exists a control $u(t)$ that will transfer the state vector from $x_o = x(t_o)$ to a predetermined value x_1 in a finite time.

---

2 This section introduces *dynamic controllability*, which is an extension of *static controllability* introduced in 1952 by Timbergen, who established that under certainty, a policy maker can make an n-dimensional system behave in the desired manner if he or she can use n instruments that directly affect every component of the state vector. For a linear system of the form: $x = Au$, static controllability is guaranteed if A is a non singular matrix. Static controllability is of little value for dynamic systems because it lacks the time element.

3 When the state vector is not directly available, controllers might want to steer the output vector (also called information vector) $y(t)$ defined in (2.1.1) from any initial value to a predetermined value y_1. When such a task is possible, the system is called *output controllable.* Output controllability is outside the scope of this book, and the reader referred to Aoki (1976) for a clear presentation of the topic.

Like observability, controllability is difficult to establish for non linear systems. It is not difficult to establish controllability for a linear system of the form (2.1.1). First, we must define the (*nxnm*) *state controllability matrix* as:

$$G = [B, AB, A^2B, A^3B, \dots A^{n-1}B]. \tag{2.1.3}$$

and use the following theorem:

Theorem: The system (2.2.1) is controllable if and only if its state controllability matrix has rank n.

Note that the definitions of controllability and observability assume that there are no constraints on time or on the possible values of the control vector. The two theorems above do not hold with constraints. For example firm's plan to gain market share might not be feasible if limitations are imposed on its advertising expenditures, or if the goal is to be reached within a very short time.

Example 1

The state controllability matrix for a system of the form (2.1.1), where $A = \begin{bmatrix} 1 & 0 & 1 \\ 0 & 1 & 1 \\ 1 & 1 & 0 \end{bmatrix}$ and $B = \begin{bmatrix} 0 \\ 0 \\ 1 \end{bmatrix}$ is: $G = \left[\begin{pmatrix} 0 \\ 0 \\ 1 \end{pmatrix} \begin{pmatrix} 1 \\ 1 \\ 0 \end{pmatrix} \begin{pmatrix} 0 \\ 1 \\ 2 \end{pmatrix} \begin{pmatrix} 2 \\ 3 \\ 2 \end{pmatrix} \cdots \right]$. The state vector has dimension equal to 3. Since $\text{rank}[B] = 1$, $\text{rank}[B, AB] = 2$, $\text{rank}[B, AB, A^2B] = 3$, the system is state controllable.

Example 2

Consider the system of example 1 where the matrix B is instead $B = \begin{bmatrix} 1 & -1 \\ -1 & 1 \\ 0 & 0 \end{bmatrix}$.

The state controllability matrix is now:

$$G = \left[\begin{pmatrix} 1 & -1 \\ -1 & 1 \\ 0 & 0 \end{pmatrix} \begin{pmatrix} 1 & -1 \\ -1 & 1 \\ 0 & 0 \end{pmatrix} \begin{pmatrix} 1 & -1 \\ -1 & 1 \\ 0 & 0 \end{pmatrix} \cdots \right].$$

Since $\text{rank}[B] = \text{rank}[B, AB] = \text{rank}[B, AB, A^2B] \dots = 1$, the system is state uncontrollable.

2.1.3 ***Robustness.***

Since models are a simplified representation of reality, it is natural to expect that very similar models will generate very similar solutions. Often, a "small" simplifying assumption has no noticeable effect on the behavior of the model. Unfortunately, it can also happen that a minuscule change in either the structure of the model or in the value of the parameters used lead to a very different system behavior. For example a physicist might be tempted to ignore the friction in modeling the behavior of a pendulum, or an economist might ignore information or transaction costs in a model. For many economic models, transaction costs do not affect the nature of the solution. However, air friction drastically changes the nature of the solution for modeling the motion of a pendulum. Without friction, the pendulum oscillates forever with constant energy, while with the friction, the energy of the pendulum converges exponentially to zero. For some models[4], a minute change in initial conditions lead to completely different behavior in the long run. Trying to explain the roots of such differences, mathematicians have developed the theories of bifurcation and chaos. These theories have been used extensively in biology to explain the cyclical explosive growth of some populations, followed by an abrupt decline[5]. These so called evolutionary models are slowly gaining acceptance in economics and finance.

For the controller, or policy maker, abrupt changes in equilibrium points or jumps in the state trajectories are undesirable. Few economists or politicians would want to use a policy that would yield the desired level of inflation only if the underlying economic model is correctly specified, and would yield a long period of hyperinflation if the model has a slight misspecification. The economic model must not only be stable in some sense (preferably exponential stability), but also must yield an acceptable result even in case the model has some specification or estimation errors. The system must be *robust.*

---

4 especially the so called "gradient models", characterized by an equation of the form: $dx/dt = -\nabla f(x)$, where $x \in \Re^n$ and ∇ denotes the gradient of f(). For several clear examples, see Beltrami (1987).

5 For example under certain conditions, algae can grow extremely fast, and cover the surface of ponds or marshes. When this occurs, fish is deprived of oxygen and dies. When the temperature drops, the algae dies, and the surviving fish stock grows back. If the fish stock is large enough to eat most of the algae, the phenomenon will not occur, and the fish stock will grow up to the point where there is not enough algae to feed them, and another cycle is started.

Definition. A system is *robust* (or *structurally stable*) if small perturbations of the system generate state trajectories that are qualitatively similar to the state trajectories of the unperturbed system.

Note that the notion of robustness does not imply any of the three kinds of stability introduced in the next section (Lyapunov, asymptotic and exponential). Robustness can be specific to estimation errors (for example, a model is robust if the parameter estimates are within two standard deviations of the true value) or to misspecification errors, in which case it is not possible to describe quantitatively the size of the "small perturbation" used in the definition. We provide some examples of robustness, and evolutionary games in the models of chapter 7.

2.2 Stability analysis

2.2.1 Definitions

As seen in the previous section, a dynamic system can be controlled only if it is observable and controllable. Since the actions of the controller are determined by the values taken by the components of the state vector, these values must be finite. Indeed, it would be impossible for a firm to maximize profits or market share if another firm was producing an infinite amount of the good, or if another firm had infinite financial resources. It would also be impossible to identify what is the optimal monetary policy if inflation or the money supply were infinite.
A system is stable in some sense if all the components of the state and control vectors are finite at all time, and the system converges to some form of steady state. There are many definitions of stability, and we present only the ones most commonly used in applied control theory; Lyapunov stability (also called "stability in the sense of Lyapunov"), asymptotic stability, and exponential stability.
Let $\dot{x} = f(x)$ represent the system of n differential equations:

$$\begin{aligned} \dot{x}_1 &= f_1(x_1, x_2, \dots x_n) \\ \dot{x}_2 &= f_2(x_1, x_2, \dots x_n) \\ &\vdots \\ \dot{x}_n &= f_n(x_1, x_2, \dots x_n), \end{aligned} \tag{2.2.1}$$

where the dot represents taking derivatives with respect to time, and $f_i(x)$ are piecewise continuous for each component of the vector $x = (x_1, x_2, \dots x_n)$, defined on a subset of R^n.

Definition: The system of differential equations (2.2.1) is *Lyapunov stable* around the point x_0 if for every $\delta > 0$, there exists a $\varepsilon(\delta) > 0$ such that for any initial point $x(t_0)$ in an ε-neighborhood of x_0, the trajectory of the solution $x(t)$ satisfies $\|x(t) - x_0\| < \delta$ for any $t > t_0$.

In other words, a point is a Lyapunov stable equilibrium if the trajectory (also called *orbit*) will not move further than δ away from that point when starting from an initial position that is sufficiently close it. Lyapunov stability is the most general definition of stability, and is what is implied by most authors when they state that a system is "stable". Following that trend, we can say that a system is unstable if it is not stable.

Asymptotic stability imposes stronger conditions that guarantee that the trajectory will not only remain in a neighborhood of the equilibrium point, but also converge to the point;

Definition: The system of differential equations (2.2.1) is *asymptotically stable* around the point x_0 if it is Lyapunov stable around that point, and there exists $r > 0$ such that $\|x(t) - x_0\| < r$ implies that $x(t)$ converges to x_0 as t becomes infinite.

The region around the equilibrium point identified by $\|x(t) - x_0\| < r$ is called the *domain of attraction.* A graphical representation of Lyapunov and asymptotic stability is given in figure 1.

In many applications, even asymptotic stability is not acceptable, because the trajectories might behave in an undesirable way during convergence to the equilibrium point. For example, if a system in a stable equilibrium is disturbed, the state vector might return to its equilibrium after going through high frequency oscillations, like a rigid metal rod hit by a light object or the string of a guitar being plucked. Exponential stability does not remove the possibility of high frequency oscillations, but provides an upper bound for the distance between the state vector and the equilibrium point that decreases with time:

Definition: The system of differential equations (2.2.1) is *exponentially stable* around the point x_0 if for any $t > 0$, there exists $\alpha > 0$ and $\beta > 0$ such that $\|x(t)\| < \alpha\|x_0\| e^{-\beta t}$ for any point in a neighborhood of x_0.

A graphical representation of the difference between asymptotic and exponential stability is given in figure 2. Clearly, exponential stability implies asymptotic stability, and asymptotic stability implies Lyapunov stability. However, asymptotic stability does not imply exponential stability, and Lyapunov stability does not imply asymptotic stability.

Two important notes: 1/ Mere convergence to a steady state is not sufficient for stability. The trajectory of the state vector must also remain close to the point of convergence. 2/ The fact that a system is unstable does not always mean that one of the components of the state vector becomes infinite. While it is true that for an unstable linear system, one of the components of the state vector becomes infinite over time, it is not a needed symptom of instability for a non linear system. Indeed, many non linear systems are unstable, yet none of the components of the state vector become infinite. The stability of non linear systems has been studied extensively and has stimulated the abundant research in bifurcation and chaos theories, as well as in evolutionary games, where a steady state might be a cycle.

The definitions of stability given above describe how the system would behave in a neighborhood of an equilibrium point. When the definitions of local stability can be extended to the entire domain of definition for the state variables instead of the neighborhood of a point, they become respectively definitions of *global stability*, *global asymptotic stability*, and *global exponential stability*. For example, a system is globally asymptotically stable around the point x_0 if the trajectory starting at any initial point will eventually converge to x_0 (i.e. the domain of attraction is the entire domain of definition of the state vector).

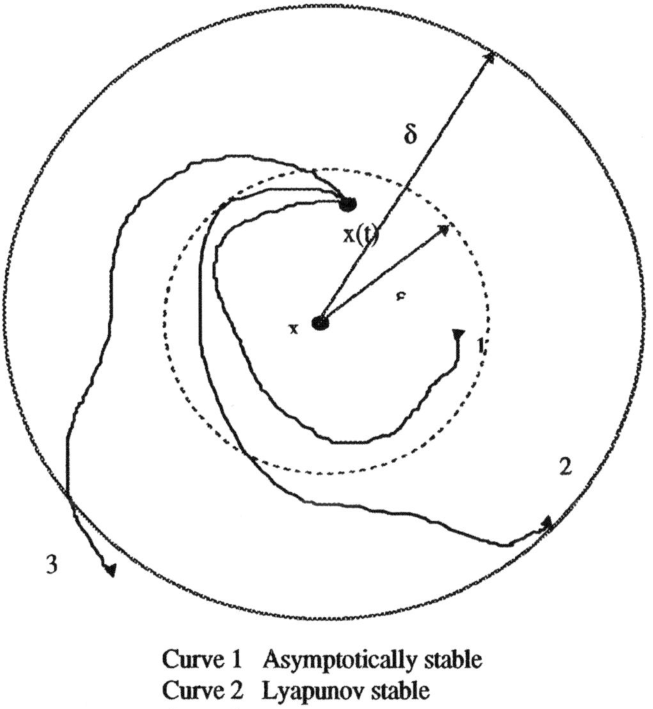

Curve 1 Asymptotically stable
Curve 2 Lyapunov stable
Curve 3 Unstable

Graphical example of asymptotic and Lyapunov stability.

Figure 1

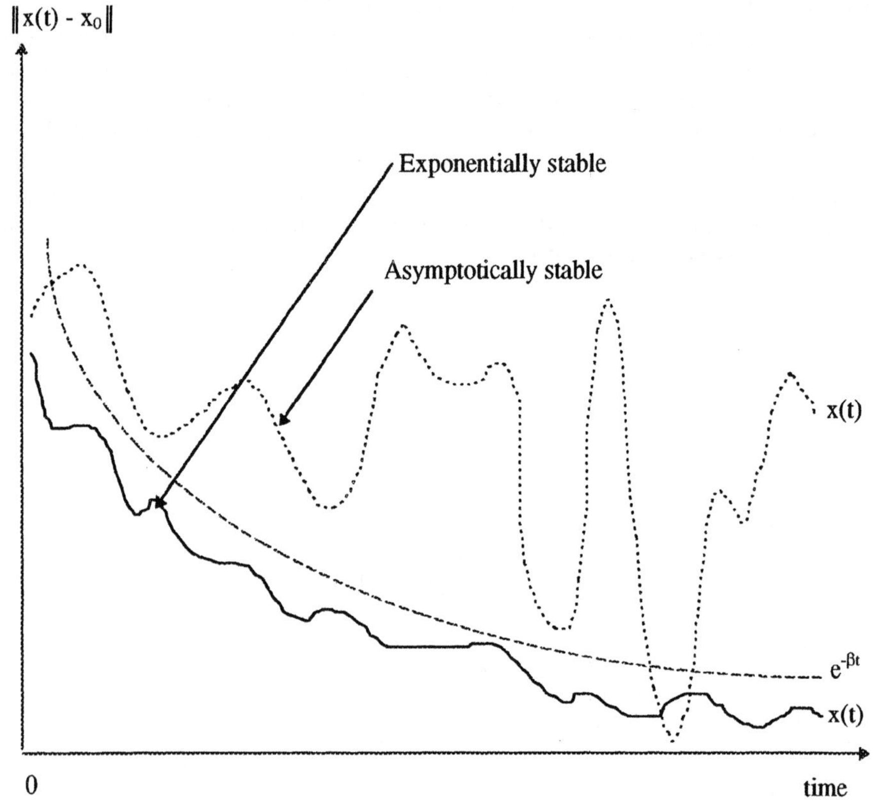

Graphical examples of stability

Figure 2

2.2.2 Stability of non linear systems.

The most common way to establish that a non linear system is stable is to linearize it using Taylor's theorem, and use results from linear systems theory. Using a first order Taylor expansion around a reference point x_0 for each equation yields the linearized system:

$$\dot{x} = J_0 x \tag{2.2.2}$$

where J_0 is the Jacobian matrix evaluated at x_0 (i.e. $j_{ij} = \partial f_i / \partial x_j |_{x=x_0}$). This powerful linearization method has its drawbacks, especially if the resulting model ignores non linear dynamics that eventually create instability. The problem can be resolved by using advanced non linear optimal control methods (for example see Slotine and Li, (1991)), or by using a *robust* suboptimal control law (with the property that small perturbations do not affect the convergence of the state vector) such as the *variable structure control* method presented in chapter 6. Many applied economists using a linearized version of a complex model obtain mediocre results, and quickly blame the results on poor estimation of the parameters, or on insufficient data, when in fact the linear structure chosen is at fault. Such problems are not encountered if the analysis is restricted to a neighborhood of the reference point used for linearization.

2.2.3 Stability of linear systems

Linear systems are either time invariant (the matrix A is independent of time) or time varying. The time invariant formulation is the most used in economics and finance for two reasons: the information necessary to estimate the parameters of time varying systems is usually not available, and the analysis needed to establish the qualitative properties of the system is much more complex than for a time invariant system. When compared with non linear systems, linear systems are very attractive because their stability, when established, is global and not just local.

If A is a non singular matrix independent of t, the time invariant linear system

$$\dot{x} = A\,x(t) \tag{2.2.3}$$

has solution:

$$x(t) = x_0\, e^{At} \tag{2.2.4}$$

where the matrix $f(A) = e^{At}$ is computed using the technique presented in the appendix. As t becomes infinite, the state trajectory will converge to a steady state (i.e. the system is globally asymptotically stable) only if the norm of the matrix e^{At} is bounded, which is guaranteed if all the eigenvalues of A have negative real parts.

One should be careful not to use the eigenvalue check instead of the norm of the state transition matrix for a time varying case. Indeed, if A is a function of time, the system is not necessarily asymptotically stable if the real parts of the eigenvalues of A are negative. Instead, since the time varying linear system:

$$\dot{x} = A(t)\,x(t) \tag{2.2.5}$$

has solution[6]

---

6 See the mathematical appendix.

$$x(t) = \left(e^{\int_{t_0}^{t} A(\tau)d\tau} \right) x_0 , \tag{2.2.6}$$

one must check that the norm of $e^{\int_{t_0}^{t} A(\tau)d\tau}$ is bounded as time becomes infinite.

Example 1

Check the stability of the system: $\dot{x} = \begin{bmatrix} -1 & -2 & -2 \\ 1 & 2 & 1 \\ -1 & -1 & 0 \end{bmatrix} x$.

The characteristic equation for A is $\det(A - \lambda I) = (\lambda-1)(\lambda^2-1) = 0$. The eigenvalues are $\lambda_1 = 1$ with multiplicity $m_1 = 2$, and $\lambda_2 = -1$ with multiplicity $m_2 = 1$. The eigenvalue test is appropriate since the system is time invariant, and we can conclude that since the eigenvalues do not all have a negative real part, the system is unstable.

Example 2

Check the stability of the system: $\dot{x} = \begin{bmatrix} -2 & -1 \\ 1 & -2 \end{bmatrix} x$.

The characteristic equation for A is $\det(A - \lambda I) = \lambda^2 + 4\lambda + 5 = 0$. The eigenvalues are $\lambda_1 = -2 + \mathrm{i}$ and $\lambda_1 = -2 - \mathrm{i}$. The eigenvalue test is appropriate since the system is time invariant, and we can conclude that since all the eigenvalues have a negative real part, the system is asymptotically stable.

Example 3

Check the stability of the system: $\dot{x} = \begin{bmatrix} -1 & e^{2t} \\ 0 & -1 \end{bmatrix} x$.

The characteristic equation for A is $\det(A - \lambda I) = (\lambda + 1)^2 = 0$. The eigenvalues are $\lambda_1 = \lambda_2 = -1$. However we can not conclude that since all the eigenvalues have a negative real part, the system is asymptotically stable. The eigenvalue test is inappropriate in this case since the system is not time invariant. Instead, we must use the method described in the appendix to compute the norm of the state transition matrix $e^{\int A(t)dt}$. After some straightforward computations, we have $e^{\int A(t)dt} = \begin{bmatrix} e^{-t} & e^{2t} \\ 0 & e^{-t} \end{bmatrix}$, which has an unbounded norm as t becomes infinite. The time varying system is therefore unstable (despite the fact that the eigenvalues of A have negative real parts)!

2.2.4 Routh-Hurwitz test

For large time invariant systems, checking the sign of the real parts of every eigenvalue of a matrix can be quite difficult as it amounts to finding the roots of a characteristic polynomial of high degree. A very elegant theorem, proven by Routh and Hurwitz, identifies the condition under which all the roots of a polynomial will have negative real parts. Because the theorem only requires the computing the leading principal minors of a matrix it is especially useful in practical applications (most computer packages cannot compute the roots of high degree polynomials). We first need the following definition;

Definition: The *Hurwitz matrix* of the polynomial with real coefficients $P(\lambda) = a_0\lambda^n + a_1\lambda^{n-1} + a_2\lambda^{n-2} + \cdots + a_i\lambda^{n-i} + \cdots + a_{n-1}\lambda + a_n$, such that $a_0 > 0$ is given by:

$$M = \begin{bmatrix} a_1 & a_3 & a_5 & \cdots & & & & & 0 & 0 & 0 \\ a_0 & a_2 & a_4 & \cdots & & & & & 0 & 0 & 0 \\ 0 & a_1 & a_3 & a_5 & \cdots & & & & 0 & 0 & 0 \\ 0 & a_0 & a_2 & a_4 & \cdots & & & & & & \\ 0 & 0 & a_1 & a_3 & a_5 & & & & \vdots & \vdots & \vdots \\ 0 & 0 & a_0 & a_2 & a_4 & \ddots & & & & & \\ 0 & 0 & 0 & a_1 & a_3 & \ddots & \ddots & & & & \\ & & & & \ddots & \ddots & \ddots & a_{n-1} & 0 & 0 & 0 \\ \vdots & \vdots & \vdots & & & \ddots & \ddots & a_{n-2} & a_n & 0 & 0 \\ & & & & & & \ddots & \ddots & a_{n-1} & 0 & 0 \\ 0 & 0 & 0 & & & & \ddots & \ddots & a_{n-2} & a_n & 0 \\ 0 & 0 & 0 & & & & \cdots & a_{n-5} & a_{n-3} & a_{n-1} & 0 \\ 0 & 0 & 0 & & & & \cdots & a_{n-6} & a_{n-4} & a_{n-2} & a_n \end{bmatrix} \quad (2.2.7)$$

The matrix is constructed as follows:

Row 1 starts with the odd coefficients $a_1, a_3, a_5, \cdots$ until all odd coefficients are used,

Row 2 starts with the even coefficients $a_0, a_2, a_4, \cdots$ until all even coefficients are used,

Rows 3 and 4 are the same as rows 1 and 2 respectively, but offset by one column,

Rows 5 and 6 are the same as rows 3 and 4 respectively, but offset by one column,

Rows 7 and 8 are the same as rows 5 and 6 respectively, but offset by one column, and so on until the main diagonal of the matrix being built contains all the coefficients of the polynomial except a_0. Finally, fill in all the unspecified elements of the matrix with zeros. The matrix should be square with dimension n (equal to the degree of the characteristic polynomial).

Theorem (Routh-Hurwitz):

The polynomial $P(\lambda) = a_0\lambda^n + a_1\lambda^{n-1} + a_2\lambda^{n-2} + \cdots + a_i\lambda^{n-i} + \cdots + a_{n-1}\lambda + a_n = 0$ with real coefficients such that $a_0 > 0$ has all its roots with negative real parts if and only if the leading principal minors of the Hurwitz matrix are all positive.

If the polynomial $P(\lambda)$ satisfies the theorem above, it is called a *Hurwitz polynomial.* Note that the theorem is valid only if the first coefficient is positive. If it is negative, simply multiply the equation $P(\lambda) = 0$ by -1. It can be shown that a polynomial is not a Hurwitz polynomial is its coefficients are not all strictly positive. If all the coefficients are strictly positive, another theorem reduces by half the number of leading principal minors to be checked.

Theorem (Liénard-Chipart)

A polynomial with real coefficients is a Hurwitz polynomial if and only if all its coefficients are positive and the Hurwitz matrix has all its leading principal minors of either odd order or even order are positive.

Example 1

Check the stability of the system: $\dot{x} = \begin{bmatrix} -1 & 0 & -1 \\ 0 & -1 & 1 \\ 1 & 0 & -2 \end{bmatrix} x$.

The characteristic equation for A is $\det(A - \lambda I) = \lambda^3 + 4\lambda^2 + 4\lambda + 1 = 0$. We have $a_0 = 1,\ a_1 = 4,\ a_2 = 4,\ a_3 = 1$. The conditions of the, Liénard-Chipart theorem are met, and we can just check the sign of the second order leading principal minor of the Hurwitz matrix;

$$M = \begin{bmatrix} a_1 & a_3 & 0 \\ a_0 & a_2 & 0 \\ 0 & a_1 & a_2 \end{bmatrix} = \begin{bmatrix} 4 & 1 & 0 \\ 1 & 4 & 0 \\ 0 & 4 & 1 \end{bmatrix}.$$

The principal minor $m_2 = \det\begin{bmatrix}4 & 1\\ 1 & 4\end{bmatrix} = 15 > 0$. Since m_2 is positive, the Liénard-Chipart theorem guarantees that m_1 and m_3 are also positive, and that the Routh-Hurwitz conditions are met. The linear system is therefore globally asymptotically stable.

Example 2

Check the stability of the system: $\dot{x} = \begin{bmatrix}1 & -2 & 0 & 0\\ 0 & 1 & 0 & -1\\ -1 & 1 & -2 & 1\\ 0 & -2 & 1 & 0\end{bmatrix}x$.

The characteristic equation for A is $\det(A - \lambda I) = \lambda^4 - 6\lambda^2 + 3\lambda = 0$. Hence, we have $a_0 = 1,\ a_1 = 0,\ a_2 = -6,\ a_3 = 3,\ a_4 = 0$. The Hurwitz matrix is the 4 by 4 matrix:

$M = \begin{bmatrix}0 & 3 & 0 & 0\\ 1 & -6 & 0 & 0\\ 0 & 0 & 3 & 0\\ 0 & 1 & -6 & 0\end{bmatrix}$. The leading principal minors are $m_1 = 0$, $m_2 = -3$, $m_3 = -9$, $m_4 = 0$. Since the principal minors are not all positive, some eigenvalues of the state transition matrix have non negative real parts, and the system is unstable.

2.2.5 *Phase diagrams.*

A *phase diagram* (also called *phase portrait*) is a graphical representation of the behavior of a system over time. Phase diagrams are extremely useful because they give some information about the qualitative properties of the system they represent without having to solve the associated differential equations. Because it is impossible to represent more than two dimensions on a flat diagram, phase diagrams can be used only to represent the behavior of two dimensional autonomous systems. Let

$$\begin{aligned}\dot{x}_1 &= f(x_1, x_2)\\ \dot{x}_2 &= g(x_1, x_2)\end{aligned} \tag{2.2.8}$$

be an autonomous system, where f and g are piecewise continuous functions, and the dot represents the derivative of x_i with respect to time. We can plot the curve representing all the combinations of points (x_1, x_2) satisfying:

$$\frac{dx_2}{dx_1} = \left(\frac{dx_2}{dt}\frac{dt}{dx_1}\right) = \frac{\dot{x}_2}{\dot{x}_1} = \frac{g(x_1, x_2)}{f(x_1, x_2)} = \alpha \tag{2.2.9}$$

where α is an arbitrary constant. Such a curve, called an *isocline*, represents the locus of points where the state trajectory has slope α. If $f(x_1, x_2) = 0$, α is infinite, and the isocline (represented by the line FF on figure 3) represents the locus of points where the state trajectory is vertical. If $g(x_1, x_2) = 0$, we have $\alpha = 0$, and the isocline (represented by the line GG on figure 4) represents the locus of points where the state trajectory is horizontal. We can now use the property that the state trajectories have the same slope along an isocline to build the phase diagram of the system. Note that unless the two axis have the same scale, the geometric slope of the trajectories will differ from the derivatives (i.e. if one inch represents either 10 x_1 or 100 x_2, a 45° line has a geometric slope of +1, but has a numeric slope $dx_2/dx_1 = 10$) The more isoclines we built, the more precise the sketch of the state trajectories.

After sketching the isoclines, we must determine how the state moves along the trajectories as time increases. For that, we only need the sign of the derivatives of x_1 and x_2 with respect to time, that is we only need to check the sign of $f(x_1, x_2)$ and $g(x_1, x_2)$. The isocline FF (along which $f(x_1, x_2) = 0$) separates the phase plane in at least two regions. In most cases, $f(x_1, x_2)$ is positive in one of the regions, and is negative in the other (although in degenerate cases, the derivative of x_1 can have the same sign in both regions). In the example of figure 3, $dx_1/dt = f(x_1, x_2)$ is negative above the isocline FF, and positive below it. Starting at any point A below the isocline FF, x_1 will increase over time, and the state trajectory will move towards the right. Similarly, starting at any point B above FF, x_1 will decrease over time, and the state trajectory will move towards the left. We can do the same kind of analysis for $\alpha = 0$ to describe the locus of points where the state trajectories are horizontal. Solving $g(x_1, x_2) = 0$ for x_2 gives the associated isocline (line GG in figure 4).

Since $dx_1/dt = 0$ on the isocline FF, and $dx_2/dt = 0$ on the isocline GG, the intersection of the two isoclines (point E) corresponds to a steady state. In other words, if the state is at $x_1 = x_1^*$, $x_2 = x_2^*$, it will remain at that point for ever. The two isoclines in the example of figure 4 divide the plane in 4 regions. Since we know the sign of dx_2/dt and dx_2/dt in each region, we can describe the direction of the state trajectories:

	dx_1/dt	dx_2/dt	Trajectory Direction
Region 1	$f(x_1, x_2) > 0$	$g(x_1, x_2) < 0$	right & down
Region 2	$f(x_1, x_2) < 0$	$g(x_1, x_2) < 0$	left & down
Region 3	$f(x_1, x_2) < 0$	$g(x_1, x_2) > 0$	left & up
Region 4	$f(x_1, x_2) > 0$	$g(x_1, x_2) > 0$	right & up

When plotting a phase diagram, it is customary to indicate the trajectory direction with two arrows at one representative point in every region, like at the points A, B, C and D on figure 4.

Now that we can identify the direction of the state trajectories around the equilibrium point, we can classify the equilibrium point (E) in one of four categories: An equilibrium point is a

- *node* if all the state trajectories converge towards the point (stable node) or away from the point (unstable node) without crossing the isoclines FF and GG.
- *saddle point* (also called *marginally stable* point) if only two state trajectories (called the *stable branches*, or *separatrix*) lead directly to the point, two state trajectories lead directly away from the point (the *unstable branches*), and all other trajectories never reach the point.
- *focus* if all state trajectories cross the isoclines FF and GG, and converge to the point (stable focus) or away from the point (unstable focus).
- *vortex* if all state trajectories are cyclical, without ever reaching the point or diverging away from it.

Stable nodes and foci are the only asymptotically stable equilibrium points of the list. As described in the previous section, the equilibrium is local if the definitions above hold only in a neighborhood of the point, and global if they hold for the entire domain of definition of the state variables. Note that in most cases, knowledge of the general direction of the state trajectories is not enough to conclude that an equilibrium point is stable or unstable.

In the preceding paragraph, we have assumed that the intersection of the two isoclines defined by $f(x_1, x_2) = 0$ and $g(x_1, x_2) = 0$ identify a single point, which we called "the" equilibrium point. In fact, it can also happen that equilibrium points of a nonlinear system form a continuous line. The equilibrium can be a limit cycle or be a degenerate equilibrium[7].

Definition: A system of differential equations has a *limit cycle* $\Gamma(x)$ if it has a non constant state trajectory $\Gamma(x)$ that is a closed curve.

Note that since the limit cycle is a closed state trajectory, as time becomes infinite, the state will assume the values of every point described by the curve an infinite amount of time, and will never stray away from the closed curve. Since limit cycles

---

[7] Some degenerate systems can also generate a closed curve as an equilibrium. However, the state trajectory of such systems can reach a point on the equilibrium curve, and remain at that point (hence the curve is not a limit cycle). For example, the system $dx_1/dt = dx_2/dt = x^2_1 + x^2_2 - 1$ is in equilibrium at any point on the unit circle. The state trajectories are straight lines with slope equal to 1. The points on the circle and below the line $x_2 = -x_1$ are stable, while the points on the circle and above the line are unstable.

are special kinds of equilibrium involving more than a single point, we must redefine what is stability for this special case:

Definition[8]: A system of differential equations is *asymptotically stable around the limit cycle* $\Gamma(x)$ if there exists $\varepsilon > 0$ such that $Lim\|x(t_0+t) - \Gamma(x)\|_\infty = 0$ as $t \to \infty$ for any initial point $x(t_0)$ satisfying $\|x(t_0) - \Gamma(x)\|_\infty < \varepsilon$.

By analogy with the classification of equilibrium points, we call a limit cycle *asymptotically stable* if all trajectories in an ε-neighborhood of it converge to it as time becomes infinite, *unstable* if the trajectories diverge from it, and *marginally stable* (equivalent of saddle point) if some trajectories converge to it while some others diverge from it.

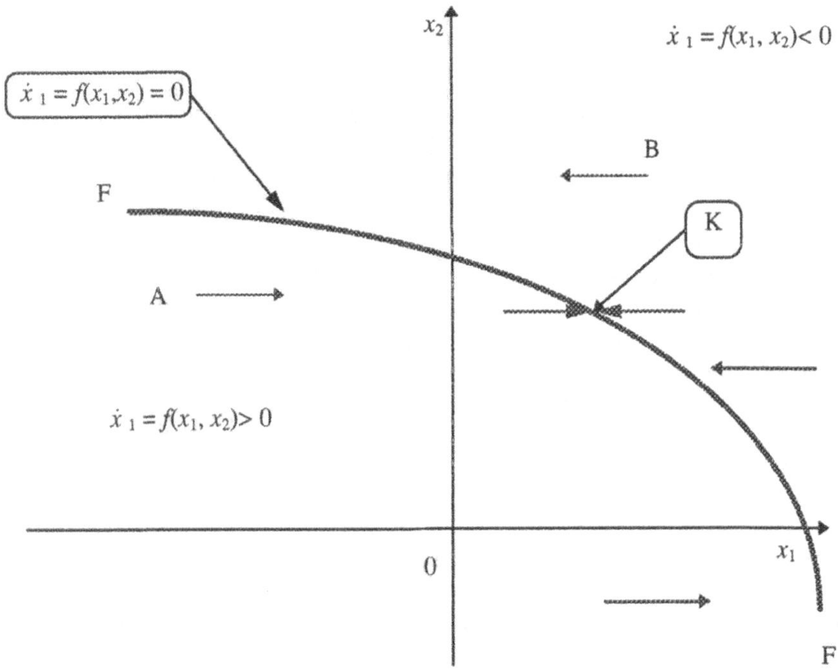

Isocline for $\dot{x}_1 = 0$

Figure 3

8 The infinity norm $\|\bullet\|_\infty$ is defined in the appendix.

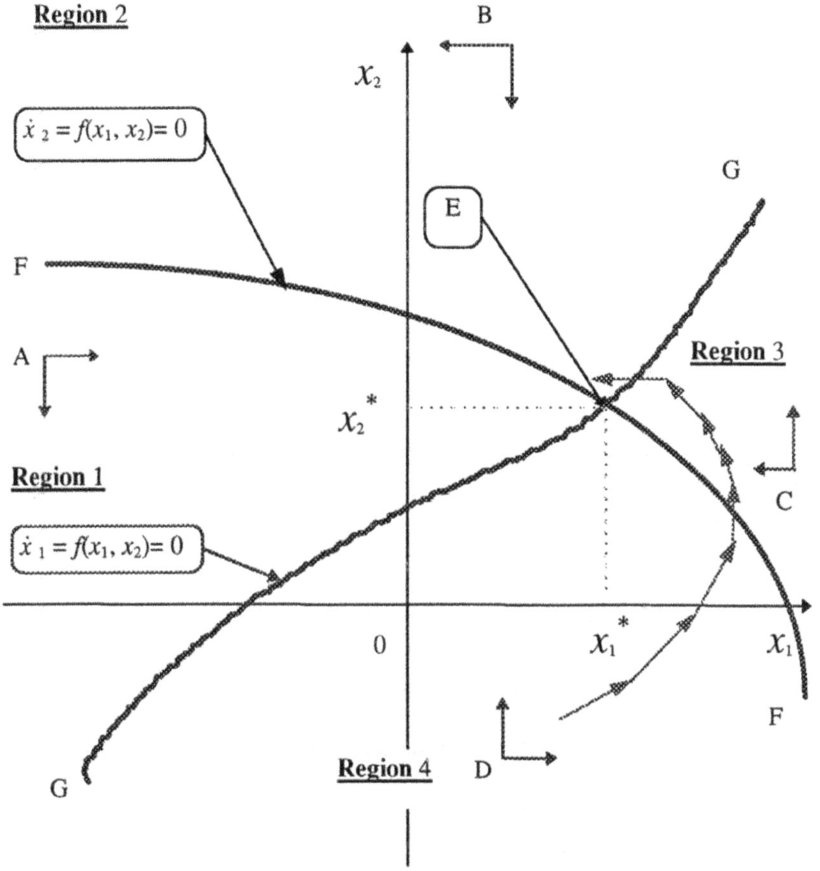

Isoclines and state trajectory (of a non linear system)

Figure 4

2.2.6 Phase diagrams of linear systems.

Linear systems have one very attractive property; If they are not degenerate (i.e. the matrix A is non singular), they have either no equilibrium point, or only one equilibrium point which is either globally exponentially stable or globally Lyapunov stable, or unstable. Linear systems of dimension 2 are extremely common in modeling, most often being the result of linearization of a non linear system around a reference point, or being a transformed second order ordinary differential equation (the origin of the system is very important because the stability properties are only local for a linearized system, and global for a transformed differential equation). The characteristic equation of second order time invariant linear systems is always of the form: $\lambda^2 + a_2 \lambda + a_1 = 0$. The shape of the phase diagrams depend upon the roots of the characteristic equation. We can write the equation as $(\lambda - \lambda_1)(\lambda - \lambda_2) = 0$ where λ_1 and λ_2 are the roots. Clearly, $a_1 = -\lambda_1\lambda_2$, and $a_2 = -(\lambda_1 + \lambda_2)$. The following covers all the possible cases;

1/ One of the roots is zero (Degenerate case: $a_1 = 0$).
The state trajectories are straight lines with slope $-a_2$ (figure 5a).

2/ The roots are real and distinct.
- if $\lambda_1 < \lambda_2 < 0$. The state trajectories are parabolas with asymptotes $x_2 = \lambda_1 x_1$ and $x_2 = \lambda_2 x_1$ (figure 5b). The state moves towards the origin, and the system is globally exponentially stable.
- if $0 < \lambda_2 < \lambda_1$. The state trajectories are parabolas with asymptotes $x_2 = \lambda_1 x_1$ and $x_2 = \lambda_2 x_1$. The state moves away from the origin, and the system is unstable.
- if $\lambda_1 < 0 < \lambda_2$. The state trajectories are hyperbolas with asymptotes $x_2 = \lambda_1 x_1$ and $x_2 = \lambda_2 x_1$ (figure 5c). The equilibrium point is a saddle point, and the system is unstable.

Note that the asymptotes are NOT the same lines as the isoclines defined by $f(x_1, x_2) = 0$ *and* $g(x_1, x_2) = 0$.

3/ The roots are real and equal: $\lambda_1 = \lambda_2 \neq 0$.
The state trajectories are parabolas with a single asymptote $x_2 = \lambda_1 x_1$.
- if the root is negative, the state move towards the origin and the system is globally exponentially stable.
- if the root is positive, the state trajectory moves away from the origin (figure 5d) and the system is unstable.

4/ The roots are purely complex: $\mathrm{Re}\,\lambda_1 = \mathrm{Re}\,\lambda_2 = 0$ ($a_1 > 0$ and $a_2 = 0$).
The state trajectories are ellipses,
- if $1 < a_1$ the ellipses are elongated along the vertical axis,
- if $0 < a_1 < 1$, the ellipses are elongated along the horizontal axis (figure 5e),
- if $a_1 = 1$, the ellipses become circles.

The system is Lyapunov stable (but not asymptotically or exponentially stable).

5/ The roots are complex with non zero real parts: $\mathrm{Re}\,\lambda_1 \neq 0$, $\mathrm{Re}\,\lambda_2 \neq 0$.
- If $\mathrm{Re}\,\lambda_1 < 0$, $\mathrm{Re}\,\lambda_2 < 0$, the state trajectories are spirals. The state moves towards the origin (figure 5f), and the system is globally exponentially stable.
- If $\mathrm{Re}\,\lambda_1 > 0$, $\mathrm{Re}\,\lambda_2 > 0$, the state trajectories are spirals. The state moves away from the origin, and the system is unstable.

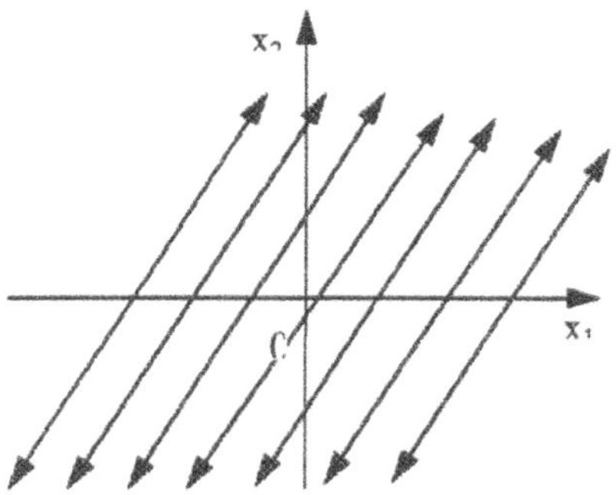

Figure 5a

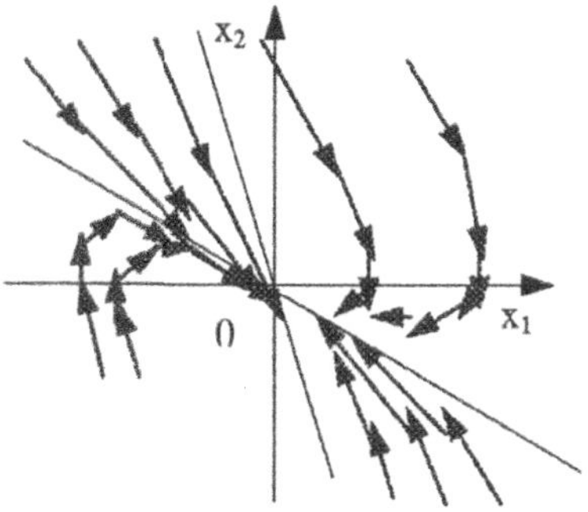

Figure 5b

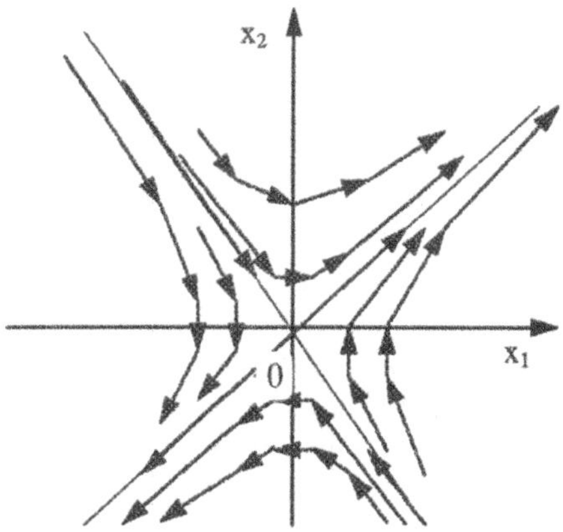

Figure 5c

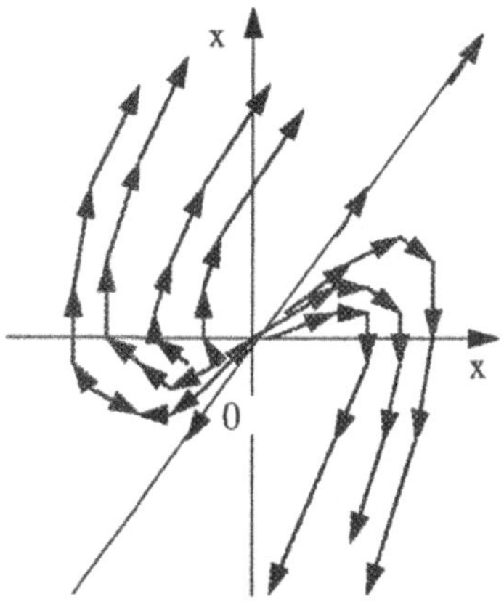

Figure 5d

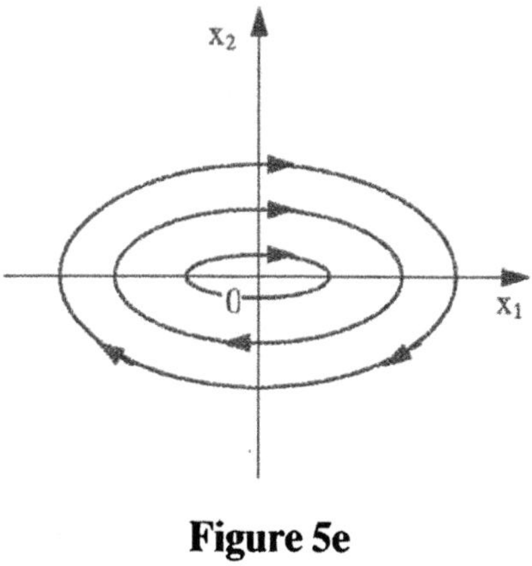

Figure 5e

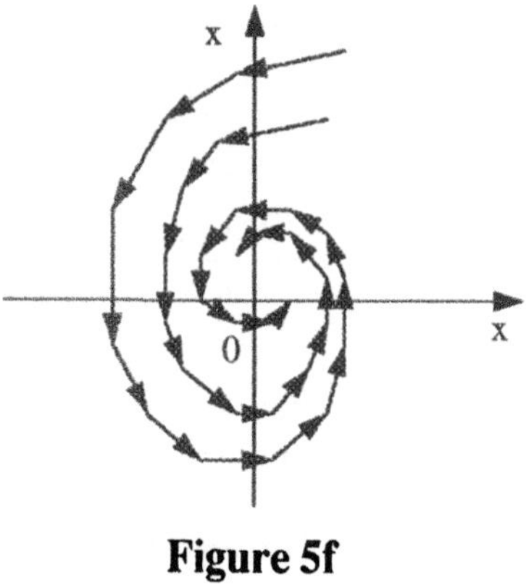

Figure 5f

Phase diagrams of linear systems

Example
Using isoclines, describe the state trajectories of the following non linear system;

$$\dot{x}_1 = f(x_1, x_2) = -2x_1 - x_2 + x_1^3 \tag{2.2.10}$$

$$\dot{x}_2 = g(x_1, x_2) = 2x_1 - x_2 \tag{2.2.11}$$

The isocline for the infinite slope is given by: $f(x_1, x_2) = -2x_1 - x_2 + x_1^3 = 0$, and the state trajectories cut vertically the isocline with equation: $x_2 = -2x_1 + x_1^3$ (I_1 on figure 6). Similarly, the isocline for the zero slope is given by: $g(x_1, x_2) = 2x_1 - x_2 = 0$, and the state trajectories cut horizontally the isocline with equation: $x_2 = -2x_1$ (I_2 on figure 6).

The two isoclines intersect at the points where $2x_1 = -2x_1 + x_1^3$. Solving for x_1 yields three equilibrium points $\alpha_0 = (0, 0)$, $\alpha_1 = (2, 4)$, and $\alpha_2 = (-2, -4)$, and the isoclines divide the plane in 6 regions. To establish the direction of the state trajectories, we first need to find the sign of the derivatives at one point not on the isoclines. Picking $A = (0, 2)$ as the representative point lying below both isoclines, we have $f(0, 2) > 0$ and $g(0, 2) > 0$. We can therefore conclude that the state trajectories lying below both isoclines are pointing to the right (since dx_1/dt is positive) and upwards (since dx_2/dt is positive). The arrows next to the representative point A for the region point up and right. There is no need to check the sign of the derivatives in any other region since the isoclines are not degenerate. Starting from the point A, the direction of the horizontal vector of a representative point is reversed every time we cross the isocline I_1 and the direction of the vertical vector is reversed every time we cross I_2.. The arrows next to the representative points A to F show the direction of the state trajectories in each region. We now have quite a bit of information about the direction of the trajectories, and yet, it is not enough to draw the trajectories accurately. Figure 6 shows a few of the possibilities, and we cannot identify which ones are correct without computing more isoclines (using (2.2.9)). We can however use linearization to identify the correct shape in a neighborhood of an arbitrary point (x_{10}, x_{20}). Since (2.2.11) is linear, there is no need to linearize it. Using Taylor's theorem, (2.2.10) can be written as:

$$\dot{x}_1 \approx f(x_1, x_2) + \left.\frac{\partial f(x_1, x_2)}{\partial x_1}\right|_{x_{10}, x_{20}} (x_1 - x_{10}) + \left.\frac{\partial f(x_1, x_2)}{\partial x_2}\right|_{x_{10}, x_{20}} (x_2 - x_{20}) \tag{2.2.12}$$

where the functions are evaluated at the point (x_{10}, x_{20}). In a neighborhood of the equilibrium point $\alpha_1 = (2,4)$, the linearized system is:

$$\dot{x}_1 = 10x_1 - x_2 - 6 \tag{2.2.13}$$

$$\dot{x}_2 = 2x_1 - x_2 . \tag{2.2.11}$$

The characteristic equation of the linearized system is: $\det(A - \lambda I) = \lambda^2 - \lambda - 8 = 0$, which has roots $\lambda_1 = 1$, and $\lambda_2 = 8$. Since the roots are both real and distinct, we can conclude that the state trajectories around the point $\alpha_1 = (2,4)$ are hyperbolas with asymptotes with slope 1 and 8, and the equilibrium point is unstable.

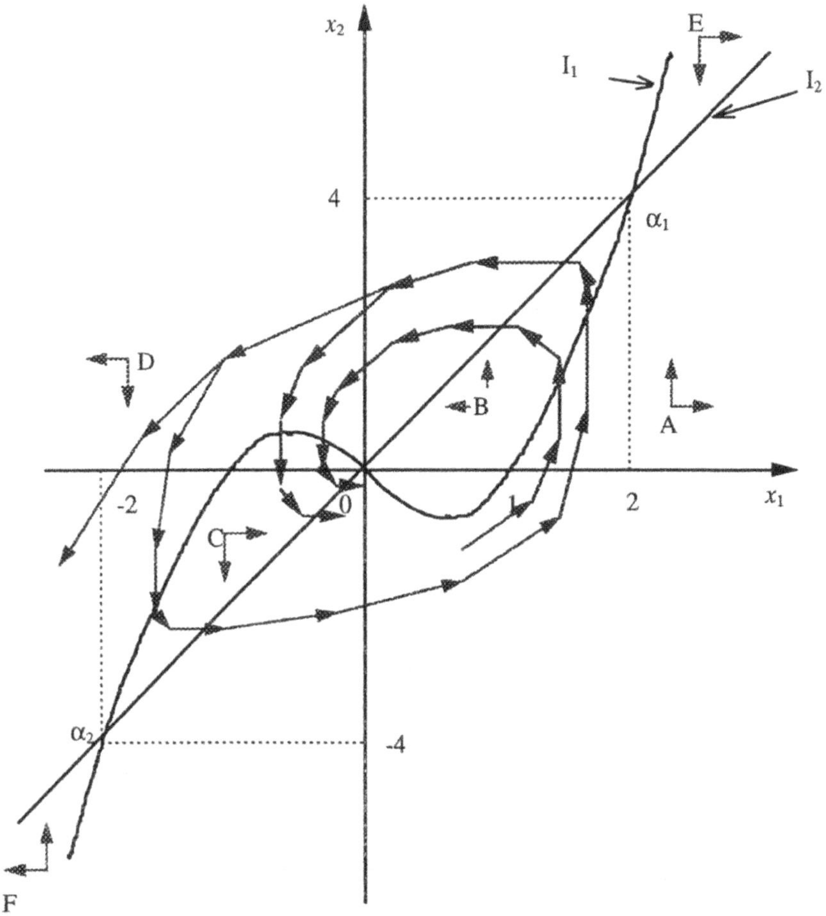

Possible state trajectories for a non linear system

Figure 6

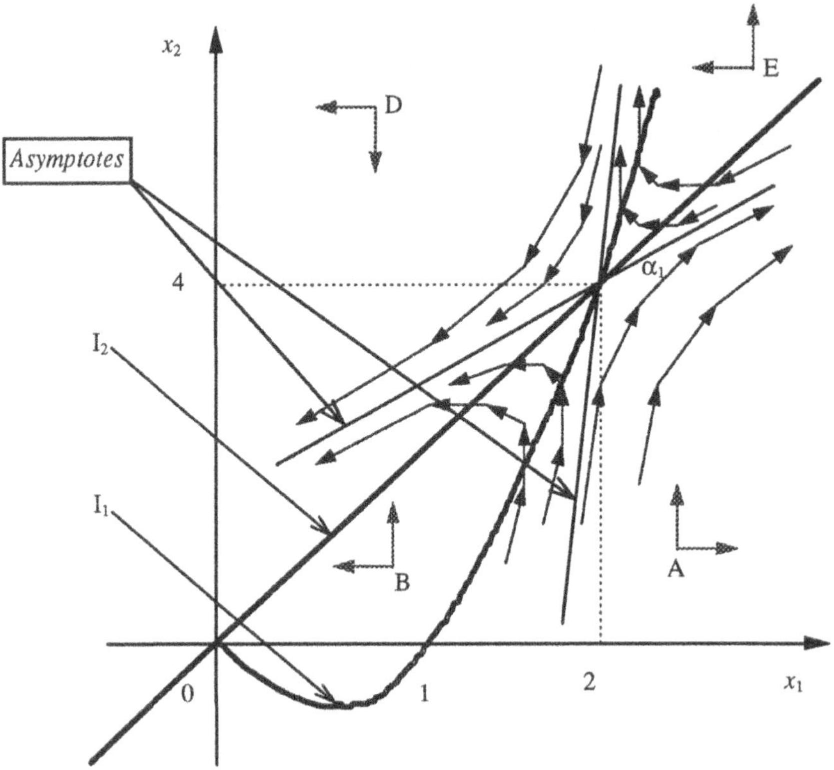

Saddle point for a non linear system

Figure 7

2.3 The maximum principle

Near the end of the 17th century, Johan Bernouilli posed what initially seemed to be a trivial problem; find the path that would minimize the time it takes a ball to roll down from one point to another. The formal answer was given decades later by the mathematician Euler, who identified the equation that must hold if the path chosen is optimal. The method by which the so called "Euler equation" was derived started an enormous amount of research in what became known as the "calculus of variations". A typical problem of the calculus of variations is to find a bounded function $I(x(t), \dot{x}(t), t)$ which is continuous in both $x(t)$ and $\dot{x}(t)$, and minimizes the definite integral $J = \int_a^b I(x, \dot{x}, t)\, dt$. The original solution of the problem was later enhanced to include constraints on the possible values $x(t)$ could take, but the initial assumption that the integrand $I(x(t), \dot{x}(t), t)$ must be continuous in $x(t)$ and $\dot{x}(t)$ could not be removed. Building on the work of several mathematicians early in this century, one American and three Russian researchers independently identified the necessary conditions to solve the more general control problem defined below, without the strict continuity assumptions for the integrand. The conditions are described in a long theorem known as the "*maximum principle*" which also allows for a wide range of constraints to be imposed on the state and the control vectors. Let us first restate the problem being solved. We are seeking the control vector $u(t) \in U \subseteq \Re^n$ that will give an extremal value to the functional

$$J = \phi(x(T),T) + \int_{t_0}^{T} I(x,u,t)\, dt \tag{2.3.1}$$

where the system is described by an n-dimensional vector $x = (x_1, x_2, \ldots x_n)$ whose components evolve according to the *state equations*

$$\dot{x}_i(t) = f_i(x(t),u(t),t) \qquad (i = 1,2,3,\ldots n) \tag{2.3.2}$$

and where the functions $I(\bullet), f_1(\bullet), f_2(\bullet), \ldots f_n(\bullet)$ are all piecewise continuous in x, u, and t. The region U is called the *admissible control region*, t_0 the starting time for the control problem, and $\phi(x(T),T)$ is a penalty function representing the cost associated with the value of the state variables at the terminal time T (the planning horizon).

The system also has known initial conditions $x(t_0) = x_0$, and must satisfy p *terminal state constraints*[9] described by:

$$\psi(x(T), T) = 0 \tag{2.3.3}$$

where ψ is a p-dimensional vector. Note that the functional ψ can represent both equality and inequality constraints, in which case the relevant "slack" variables must be introduced.

Before we state the conditions identifying the optimal solution of the control problem, it is convenient to define a functional, called the *Hamiltonian*, which includes both the integrand and the constraints (on the system and on the control) at

9 Note that the functions $\phi(x(T),T)$ and $\psi(x(T),T) = 0$ are not the same! $\phi(x(T),T)$ is a scalar which represents the cost associated with the terminal condition $x(T)$, while $\psi(x(T),T) = 0$ is a p-dimensional vector of functions describing the condition that must hold at the terminal time.

any point in time. For the system described by (2.3.1) and (2.3.2), the Hamiltonian functional is:

$$H(x,u,\mu,t) = I(x,u,t) + \sum_{i=1}^{n} \mu_i(t) f_i(x(t),u(t),t). \tag{2.3.4}$$

The additional variables introduced, $\mu_i(t)$ are called *costate* (or *adjoint*) variables, and are similar to time-dependent Lagrange multipliers. Using the more compact vector notation $f(x,u,t) = (f_1(x,u,t), f_2(x,u,t), \ldots f_n(x,u,t))'$ and $\mu(t) = (\mu_1(t), \mu_2(t), \ldots \mu_n(t))'$, the state equations and the Hamiltonian can be written as:

$$\dot{x}(t) = f(x,u,t) \tag{2.3.5}$$

and

$$H(x,u,\mu,t) = I(x,u,t) + \mu'(t) f(x,u,t)\,. \tag{2.3.6}$$

2.3.1 Conditions of the maximum principle.

The conditions that must be satisfied by the optimal control vector were derived by Pontryagin and his co-authors in a 40 page proof which requires an advanced knowledge of calculus. Because it is so long and involved, the proof is left out of this book[10]. The end result is deceivingly simple; For a control vector $u^*(t)$ to be optimal, the following conditions must hold:

$$H(x^*,u^*,\mu^*,t) \le H(x^*,u,\mu^*,t), \qquad u(t) \in U \subseteq \Re^n \tag{2.3.7}$$

$$\dot{x} = \frac{\partial H}{\partial \mu}, \tag{2.3.8}$$

$$\dot{\mu} = -\frac{\partial H}{\partial x}, \tag{2.3.9}$$

$$(\phi_x + \psi'_x \nu + \mu)'\big|_{t=T}\, dx(T) + (\phi_t + \psi'_t \nu + H)'\big|_{t=T}\, dT = 0, \tag{2.3.10}$$

where x^* and μ^* are optimal values of the state and costate vectors, the subscripts represent partial derivatives, and $\nu = (\nu_1, \nu_2, \ldots \nu_p)'$ is a vector of Lagrange multipliers associated with the p terminal constraints (2.3.3). Equation (2.3.10) is known as the *transversality conditions*, and is derived with Leibnitz's rule, and

10 See Burghes and Graham (1980) for a clear intuitive verification of the principle. In their book, they present the case where the state and control functions are both continuous as an extension of results from the calculus of variations.

using the variational calculus assumptions that T is finite[11], and that $H(x,u,\mu,t)$ is piecewise continuously differentiable in x in a neighborhood of T. It is important to note the following points about the system (2.3.7-10);

1/ The inequality (2.3.7) must hold for the optimal values of x^* and $u^* \in U$, and not just for $H(x^*,u^*,\mu^*,t) \leq H(x,u,\mu,t)$.

2/ Since the Hamiltonian is not necessarily differentiable, or even continuous in $u(t)$, the relation (2.3.7) can be replaced by $\frac{\partial H}{\partial u} = 0$ only when $H(x,u,\mu,t)$ is differentiable with respect to u. In such case, the second order conditions should also be checked to identify whether the solution corresponds to a maximum or a minimum of H.

3/ The costate vector $\mu(t)$, whose evolution is given by the costate equation (2.3.9), is a function of only one variable; time.

4/ For μ to be defined, H must be piecewise continuously differentiable in x.

5/ At any point in time, the elements of the costate vector represent the shadow price of the corresponding elements of the state vector.

6/ The vector v is a vector of scalar Lagrange multipliers independent of time since it applies only to the terminal constraint at time T and does not appear in either the state or costate equations.

The conditions of the maximum principle are usually not sufficient to fully identify an optimal control vector, and without further restrictions, there is no guarantee that the control vector is unique. Two theorems spell out when the conditions of the maximum principle are not only necessary, but also sufficient;

Theorem (Mangasarian)
The conditions of the maximum principle are sufficient for the maximization of $J = \int_{t \in \Omega} I(x,u,t)\, dt$ if the functions $I(x,u,t)$, $f_1(x,u,t)$, ... $f_n(x,u,t)$ are differentiable and concave in x and u, and if $\mu_i \geq 0$ for $u_i = u_i^*(t)$ and for any $\Omega = [t_0, T]$ where T is fixed.

Theorem (Arrow)
The conditions of the maximum principle are sufficient for the maximization of $J = \int_{t \in \Omega} I(x,u,t)\, dt$ if the functional $H(x,\mu,t) = I(x,u^*,t) + \sum_{i=1}^{n} \mu_i(t) f_i(x(t),u^*(t),t)$ is concave in $x(t)$ for any $t \in [t_0, T]$, where $u^*(t)$ maximizes $H(x,u,\mu,t)$ and where T is fixed.

11 The transversality conditions are not necessary for optimality when T is infinite.

The functional $H(x,\mu,t)$ is called the *Maximized Hamiltonian function*, and differs from the Hamiltonian defined by (2.3.4) because the function $u^*(t)$ has been substituted in the Hamiltonian which is now a function of the three variables x, μ, and t.

2.3.2 *Boundary Conditions*

The conditions of the maximum principle identify two systems of first order differential equations, which can be solved explicitly if the initial vectors $x(t_0)$ and $\mu(t_0)$ are known. Unfortunately, the initial value of the costate vector is unknown. When the state and costate equations form a system of coupled differential equations (x is a function of μ, and μ is a function of x), the system can be solved if any two boundary values are given. The general method of solution is to solve the system of differential equations defined by the maximum principle, and then to identify the coefficients with the help of the initial and terminal state conditions. The functional $\psi(x(T),T)$ takes a specific form for each possible terminal constraint;

Case 1: Fixed terminal time;
If $x(T)$ is free and T is fixed, (2.3.3) reduces to: $T\text{-}T_f = 0$, where T_f is the desired terminal time. Since the terminal time is fixed, $dT = 0$, and (2.3.10) reduces to:

$$(\phi_x + \psi'_x\, \nu + \mu)'\Big|_{t=T} dx(T) = 0. \tag{2.3.11}$$

Case 2: Fixed terminal state;
If $x(T)$ is fixed and T is free, (2.3.3) reduces to: $x(T) - x_T = 0$, where x_T is the desired value of the state vector at the end of the planning horizon. Since the final value of the state variable is fixed, $dx(T) = 0$, and (2.3.10) reduces to:

$$(\phi_t + \psi'_t\, \nu + H)'\Big|_{t=T} dT = 0, \tag{2.3.12}$$

Case 3: Fixed terminal curve;
If $x_T = x(T)$ is a function of the terminal time T, the final value of the state variable is now dependent upon T, and we have $dx(T)/dT \neq 0$, and (2.3.10) applies.

Case 4: Fixed terminal point;
If both $x(T)$ and the terminal time T are fixed, we have $dx(T) = 0$ and $dT = 0$. Equation (2.3.10) does not apply since the fixed terminal state and fixed terminal time are sufficient to identify the parameters of the system of differential equations.

Some problems are a combination of cases 1 to 4. However that the mix of initial values, terminal values and terminal constraints is of no importance, as long as there are enough restrictions to fully identify the system (for example, the Hamiltonian system with fixed final time is a system of differential equations of order $2n$, which can be identified with n initial values $x(t_0)$, p terminal constraints $\psi(x(T),T) = 0$, and n-p boundary conditions $(\phi_t + \psi'_t\, \nu + H)'\Big|_{t=T} dT = 0)$.

Numerical solution of the finite horizon problem:

Unless the system has a very simple structure and dynamics, an analytic solution of (2.3.7-10) is extremely difficult to obtain. However numerous methods are available to solve the system numerically, especially for the case where the control $u(t)$ can be expressed as an explicit function of the state and costate variables. In such case, $u(t)$ can be substituted into the Hamiltonian system (2.3.8-2.3.9), which is subject to the final condition (2.3.10) and into the final constraint (2.3.3). The most successful numerical method to solve the fixed terminal time problem is to follow the sequence:

1. guess an initial value for the costate vector $\mu(t_0)$.
2. integrate numerically the Hamiltonian system to yield the estimated values $x^e(T)$ and $\mu^e(T)$.
3. Evaluate $\psi(x^e(T),T)$ and $\phi(x^e(T),T)$.
4. Seek a vector ν^e of Lagrange multipliers that will satisfy (2.3.10)
5. If no such vector exists, then compute the *sensitivity matrix*[12]:

$$\Theta = \left[\frac{\partial x(T)}{\partial \mu(t_0)} \quad \frac{\partial \mu(T)}{\partial \mu(t_0)} \right]'$$

6. Compute the desired changes $\Delta x^e(T)$ and $\Delta\mu^e(T)$ that would make $x^e(T) + \Delta x^e(T)$ and $\mu^e(T) + \Delta\mu^e(T)$ satisfy (2.3.3) and (2.3.10).
7. Estimate the change needed in the initial guess of $\mu(t_0)$, using the relationship $[\Delta x^e(T), \Delta\mu^e(T)]' = \Theta\, \Delta\mu(t_0)$.
8. Repeat the process for the new initial guess $\mu(t_0)+\Delta\mu(t_0)$ until (2.3.3) and (2.3.10) are satisfied with an acceptable error level.

---

12 If the system is time invariant and linear (e.g. $dx/dt = Ax + Bu$), the sensitivity matrix can be r computed by integrating the Hamiltonian system n times; If e_i denotes the ith column of the nxn identity matrix, the rows of the sensitivity matrix are the vectors $(x^i(T), \mu^i(T))'$ where $x^i(T)$ and $\mu^i(T)$ are the final values of the system with initial conditions $(x(t_0), e_i)$. For a clear example, see Lewis, p177-179.

2.3.3 Infinite horizon.

If the planning horizon is infinite, the controller does not know or cannot assign the value of the state vector at the terminal time. Yet, he or she can make the assumption that the system has reached a "steady state" which can be either an unknown constant value, or a cycle. The kind of steady state depends upon the system dynamics, the objective function and constraints on the state and control vectors.

The results obtained for the finite horizon cases do not extend to the infinite horizon, in great part because the transversality conditions (that were derived using some of the assumptions of the calculus of variations) do not extend to the infinite horizon. In a paper still debated 20 years later, Halkin (1974) has shown that the conditions of the maximum principle are necessary for piecewise optimality[13] of an infinite horizon problem, with the exception of the transversality conditions. Some results have been reached in the special case where the partial derivative of the integrand of the objective function with respect to the state vector is bounded (Seierstadt and Sydsæter (1987) (chapter 3-7)) or when the integrand is independent of time (Michel (1982)). However, extending the transversality conditions to the infinite horizon usually leads to faulty results[14] (especially when the control or the state variables are bounded).

Infinite horizon problems have been used because the mathematical formulation (and solution) is often less complex than the formulation of finite horizon problems. However, the adoption of an infinite horizon has been criticized on two major points;

1/ Irrelevance of the assumptions and time inconsistencies:

Economic growth models often assume that since an economy will last as long as the world will exist, the relevant horizon is infinite. However the choice of an infinite horizon can yield impractical short term solutions as the state trajectories might converge to some steady state at a time so far in the future that the economic assumptions made to formulate the problem might then be obsolete Even with a finite horizon, the time scale might create some inconsistencies between the "long run" and the "short run" optimal solutions (for example firms must balance their long run objective to gain market share, with the short term profit maximization).

2/ Implied convergence of the objective function.

Most of the infinite horizon models proposed are solved with the help of the transversality conditions. However, the assumption that the state vector will assume a steady state is often made for convenience (to allow for an explicit solution of the model), and alternative outcomes are not considered. The assumption is however gradually being discarded, mostly by European economists who are more familiar with the work of Richard Goodwin, and models generating stable limit cycles are

---

13 The pair $\{x^*, u^*\}$ is *piecewise optimal* if it is optimal for the finite time problem where $t \in [t_0, T]$ with terminal constraint that $x(T) = x^*(T)$.

14 For some examples, see Halkin (1974), and Arrow & Kurtz (1970).

now appearing not only in the theory of business cycles but also in natural resources, and finance (Dockner and Feichtinger (1991), Chiarella (1990), Wirl (1994) Goodwin (1991)).

2.4 Constraints

2.4.1 Constraints on the state, control and information vectors

The constraints imposed on a system can be classified in three large groups; The first group is the set of terminal constraints that are imposed at the terminal time. Such constraints include restrictions on time, or on the terminal value of the state vector, or both. Such restrictions are captured by the function $\psi(x(T), T) = 0$ introduced in the preceding section. The second group includes constraints on the state or control or information vectors individually. The last group includes interdependent constraints on the state and control vectors. A common problem is when the maximum value of the control (e.g. prime rate) is a function of the state (e.g. unemployment rate). In all cases, the constraints can be either equality or inequality constraints (which are reduced to equality constraints with the help of slack variables), and the problem is solved with the Kuhn-Tucker approach.

2.4.2 saturation, hysteresis, discontinuities

When components of the control vector reach the boundary of the admissible control region, the control vector is called *saturated.* The solution of an optimal control problem obtained using the calculus of variations approach (the Euler Equations) is the same as the solution obtained using the maximum principle in the case where the Hamiltonian is piecewise continuously differentiable in $x(t)$ and $u(t)$. However, unlike problems that can be solved with the calculus of variations, the maximum principle can yield an optimal control vector with components that switch abruptly from one value to another. In other cases, the system dynamics is different when the value of a component is increased or decreased. The asymmetry of dynamics, called *hysteresis,* is common when the Hamiltonian includes the absolute value of components of the state or control vectors. A classic example of an economic model with hysteresis is the Phillips relation between expected inflation and unemployment. A multitude of problems can be caused by discontinuities in the control vector, or by jumps in the objective function when the jumps are caused by the number of switches in a discontinuous control, and not only by the value of the control. In the most extreme cases, the Hamiltonian function can b piecewise continuous in the state variable, but nowhere differentiable.

2.4.3 Switching curves, Bang-bang and Bang-off-bang controls

The Hamiltonian function given by (2.3.6) can also be written as:

$$H(x,u,\mu,t) = H_0(x,\mu,t) + H_1(x,u,\mu,t)\,. \tag{2.4.1}$$

where H_0 is independent of the control, and every term in H_1 is a function of the control. To satisfy conditions of the maximum principle, we must seek the vector $u^*(t)$ satisfying (2.3.7). The optimal vector $u^*(t)$ is therefore the vector that will make H_1 as negative as possible. The simplest case occurs when the implicit function theorem allows H_1 to be written as:

$$H_1(x,u,\mu,t) = g(u)\, S(x,\mu,t)\,. \tag{2.4.2}$$

(i.e. H_1 is separable in $u(t)$). Let $\mathbb{U}_{max}$ and $\mathbb{U}_{min}$ denote the set of all the values of $u(t)$ in the admissible region that correspond to the maximum and the minimum of $g(u)$ respectively. The optimal control vector is:

$$\begin{aligned} u(t) &\in \mathbb{U}_{max} && \text{if } S(x,\mu,t) < 0 \\ u(t) &\in \mathbb{U}_{min} && \text{if } S(x,\mu,t) > 0 \end{aligned} \tag{2.4.3}$$

and $u(t)$ is indeterminate if $S(x,\mu,t) = 0$. The special case where $g(u) = u$ was introduced by Pontryagin et al. (1962) as an example of how the maximum principle can be applied to problem that are not suited for the calculus of variations[15]. Note that the optimal control vector is usually not unique; For example, if $g(u) = (u_1^2 + u_2^2)$, and the admissible region for the 2-dimensional control vector is defined by $u_1 \in [-1,1]$, and $u_2 \in [-1,1]$, u_{max} can take 4 possible values, while there is only one possible value for u_{min}. For obvious reasons, the function $S(x,\mu,t)$ is called the *switching function*, and the control is called *bang-bang*. A similar type of optimal control is possible when the control vector has two sets of constraints (for example a constraint $|u(t)| < k_1$, and an integral constraint of the form $\int_{t\in[0,T]} u(t)dt) < k_2$). Such controls are called *bang-off-bang* to denote the fact that it takes two possible values (one each in $\mathbb{U}_{max}$ and $\mathbb{U}_{min}$), and that the integral constraint might yield a period when it is optimal not to control the system. The voluminous literature on "turnpike theorems" provides numerous examples of such controls. An economics example is provided by the problem faced by a firm who wants to maximize its market share through advertising, and with a limited advertising budget for the period considered. One of the possible solutions yields alternate periods of advertising and no advertising.

15 For a detailed and clear two-dimensional example of a *bang-bang* control, see Lewis (1986), pages 262-267.

2.5 Example 1; A limit pricing model

Following Fanchon, Rifkin and Sengupta (1987), we consider the output strategy of a dominant firm or a cartel attempting to maximize the present value of its stream of profits from a market with potential entry. The rate of entry into the industry is assumed to be a function of the current price only and entry is defined as an increase in output from competitors which might or might not be already in the market. In the dominant firm model, the dominant firm quotes the market price and competitors supply all they want at that price. The dominant firm then supplies the demand not met by rivals at the given price. If the barriers to entry are great, the dominant firm can produce the short run profit maximizing level of output with little fear of losing its market share. However if entry is relatively easy, the firm can increase its output to the point where the price corresponding to aggregate industry supply induces no entry. This price has been defined as the limit price. If the demand or the production of rivals is not known precisely, the dominant firm must adjust its output in order to maintain the quoted market price. While this classic formulation leads to high uncertainty about the planned level of output, and no uncertainty about the price, such a formulation can be very costly to the firm operating in a region of steep marginal cost or if the production process takes a significant amount of time. Since the price is ultimately determined by the market, and not by the firm, price is the wrong control variable.

If the market allows small fluctuations about the quoted price, the dominant firm has more room for error in determining the optimal price and the most relevant demand function is then of the form: $q=f(p,e,t)$ where e is an error term. The dominant firm can then quote its expected price and plan output at the same time, hoping that after the market makes the final price adjustment, actual price and expected price will coincide.

This model departs from the now traditional formulation of dynamic limit pricing on two points: First; it is assumed that the dominant firm has imperfect knowledge of the parameters of the market and determines its planned output as a function of expected price. Second; it is assumed that the inverse demand function contains a term reflecting expected market growth. The inverse demand function also contains an additive error term, with known stochastic properties and independent of time, representing uncertainty about the aggregate demand at any price.

A Stochastic Market Growth Model

The suppliers of a commodity are divided in two groups; the dominant firm or a cartel (also called the "dominant" firm in this paper), and other suppliers called the rivals (or "fringe"). The equilibrium market price of the product at time t is determined by the level of output of the firm and of rivals or entrants, and grows exponentially. The market demand function is assumed to be of the form:

$$p(t) = ae^{nt} - b_1(q(t)) + x(t)) + u \tag{2.5.1}$$

where $p(t)$ is the price of the commodity, $q(t)$ is the output of the dominant firms, $x(t)$ is the output of the "fringe.", a and b_1 are positive constants, n is the growth rate of demand, and u is an error term independently distributed with known mean zero, and variance v.

The dominant firm estimates that the rate of entry (or exit) of new firms is defined by:

$$x(t) = k(\text{E}[p(t)] - \mathbf{p})\,, \qquad x(0) = x_0 \tag{2.5.2}$$

where k is a positive response coefficient which reflects in some sense the willingness of the dominant firm to give the newcomers any part of the market (the smaller the value of k, the greater the risk of failure for a potential entrant). **p** is the (fixed) limit price defined as the price at which net entry equals 0, and E[.] is the expectation operator

The objective of the dominant firms is to maximize the present value of the utility of expected profits given by;

$$U(\text{E}[\pi(t)]) = \text{E}[\pi(t)] - m\ \text{var}(\pi\,(t)) \tag{2.5.3}$$

Where m is a risk aversion parameter, c is the average total cost of production of the dominant firm, assumed to be constant over time, and

$$\text{E}[\pi(t)] = (\text{E}[p(t)] - c)\ q(t) \tag{2.5.4}$$

$$\text{Var}\ \pi(t) = vq^2(t) \tag{2.5.5}$$

Note that since the dominant firm is producing in a region of constant average cost, the firm can capture most of the increase in demand with no change in price. The limit price in this model is therefore assumed constant (the results derived are no longer true if the dominant firm is operating in a region of non constant average production cost).

The optimal output strategy of the dominant firm, and the corresponding price path, is thereby obtained by maximizing the expected utility of the net payoff J over a infinite horizon. Hence the objective function of the dominant firm is:

$$J = \int_0^\infty e^{-rt}[(a^{nt} - b_1q(\text{t}) - b_1x(t) - c)q(t) - mvq^2(t)]dt$$

i.e.

$$J = \int_0^{\infty} e^{-rt}[ae^{nt} - b_1x(t) - b_2q(t) - c]q(t)dt \tag{2.5.6}$$

where r is the discount rate of the dominant firm (assumed exogenous) and

$$b_2 = mv + b_1. \tag{2.5.7}$$

The Hamiltonian of the problem and the corresponding optimizing conditions are:

$$H = e^{-rt}(ae^{nt} - b_1x(t) - b_2q(t) - c)q(t) + \lambda(t)k(ae^{nt} - b_1x(t) - b_1q(t) - \mathbf{p}) \tag{2.5.8}$$

where $\lambda(t)$ is the costate variable which represents the "optimal" shadow price of entry to the dominant firm. Assuming an interior solution, and after we make the transformation $\lambda(t) = e^{-rt}z(t)$, an optimal output strategy is defined by:

$$\frac{\partial H(x^*, z^*, q^*)}{\partial q^*(t)} = e^{-rt}(ae^{nt} - b_1x(t) - 2b_2q((t) - c) - e^{-rt}z(t)kb_1 = 0 \tag{2.5.9}$$

$$\dot{x}^*(t) = k(ae^{nt} - b_1x^*(t) - b_1q^*((t) - p), \tag{2.5.10}$$

$$\dot{z}^*(t) = b_1q^*(t) + (b_1k+r)z^*(t) \tag{2.5.11}$$

$$\lim_{t\to\infty} z^*(t) = K, \qquad \text{where } K \text{ is a constant.} \tag{2.5.12}$$

The change of variables: $X(t) = x(t)e^{-nt}$, $Q(t) = q(t)e^{-nt}$, and $Z(t) = z(t)e^{-nt}$ reduces the system to:

$$\dot{X}^*(t) = -(kb_1 + n - kA_0)\, X^*(t) + k^2A_0\, Z^*(t) + A_1(t) \tag{2.5.13}$$

$$\dot{Z}^*(t) = -A_0\, X^*(t) + (kb_1 + r - n - kA_0)\, Z^*(t) + A_2(t) \tag{2.5.14}$$

where $A_0 = b_1^2/2b_2$, $A_1(t) = k(a - aA_0/b_1 - (p + cA_0/b_1)e^{-nt})$, and $A_2(t) = (A_0/b_1)(a - ce^{-nt})$. The characteristic equation of the system and its roots are:

$$f(\theta) = \theta^2 + (2n - r)\theta + (n^2 - nr - b_1kr - b_1^2k^2 + rkA_0 + 2k^2b_1^2A_0) \tag{2.5.15}$$

$$\theta_1,\ \theta_2 = -\tfrac{1}{2}(2n - r \pm d^{1/2}) \tag{2.5.16}$$

where: $d = 4(b_1kr + b_1^2k^2 - rkA_0 - 2b_1k^2A_0) + r^2$

PROPOSITION 1; The optimal level of output $q^*(t)$ is always above the short run payoff maximizing level of output $q_o(t)$ at every point along the optimal trajectory.

(i.e. if the dominant firm were to abandon the optimal policy and revert to short run profit maximizing, it would always reduce output, and both price and entry would increase).

Proof: Assuming that the dominant firm has been moving along the optimal trajectory, at time t the short-run payoff is $h(q) = [ae^{nt} - b_1x^*(t) - b_2q(t) - c]\, q(t)$, where the asterix denotes that $x(t)$ is on the optimal path. Maximizing this short-run payoff $h(q)$ leads to the myopic optimal output rule: $q^o(t) = (2b_2)^{-1}(ae^{nt} - b_1x^*(t) - c)$. Since the long run optimal level of output $q^*(t)$, must also maximize the Hamiltonian at every point on the optimal path, equation (2.5.8) implies: $q^*(t) = (2b_2)^{-1}[ae^{nt} - b_1x^*(t) - c - kb_1z^*(t)]$ where $z^*(t)$, the shadow price of entry, is necessarily a negative quantity. Since k and b_1 are positive parameters, we have: $q^*(t) > q^o(t)$. QED.

Note that the expressions for the short run and long run output rules yield:

$$\partial q^o(t)/\partial m = -q^o(t)(v/2b_2) < 0$$

$$\partial q^*(t)/\partial m = -q^*(t)(v/2b_2) < 0$$

where the risk aversion parameter (m) and the variance of the error term (v) are defined in (2.5.3), 2.6.5) and (2.5.7). Since prices are inversely related to output, the myopic price $p^o(t)$ is higher than the long run optimal price $p^*(t)$. Hence we have: $p^o(t)/m > 0$ and $p^*(t)/m > 0$ which implies:

PROPOSITION 2; An increase in risk aversion leads to a decrease in myopic and long run output, with a corresponding increase in price.

PROPOSITION 3; Under suitable conditions, an increase in risk aversion or greater uncertainty about the demand accelerates the convergence of the price-output vector to the steady state.

Proof: The general solution of (2.5.13-14) is;

$$X^*(t) = C_1e^{\theta 1} + C_2e^{\theta 2} + \alpha_0(t)$$

$$Z^*(t) = \alpha_1e^{\theta 1t} + \alpha_2e^{\theta 2t} + \alpha_3(t)$$

where the α_i's are known functions of the given parameters of the problem, and C_i's are suitable functions determined by the boundary conditions. From (2.5.16) we have;

$$\lim_{b2 \to \infty} \theta_i = -\tfrac{1}{2}(2n - r + (2b_1k \pm r))$$

and the largest root, say θ_1, approaches $(b_1k + r - n)$ as $t \to \infty$. Hence whenever the response coefficient k satisfies $kb_1 > (n - r)$, increasing b_2 will eventually yield a positive root. In this case, the system has a stable solution only if the initial conditions guarantee that $C_1 = 0$ and α_3 is bounded (the transversality condition (12) is met only if $C_1 = 0$ and α_3 is bounded as $t \to \infty$). The preceding system reduces to:

$$X^*(t) = C_2e^{\theta 2t} + \alpha_0(t)$$

$$Z^*(t) = 2C_2e^{\theta 2t} + \alpha_3(t)$$

where $\alpha_0(t)$ and $\alpha_3(t)$ tend to suitable constants as t becomes infinitely large. Since the discriminant of the characteristic equation is positive, an increase in b_2 leads to a higher absolute value of θ_2 (which is negative) and therefore increases the rate of convergence to the steady state. QED.

Stability conditions and long run analysis.
After the change of variables, the reduced form of the system (2.5.13-14) is:

$$\dot{X}^*(t) = -(kb_1 + n)\, X^*(t) - kb_1Q^*(t) + k(a\text{-pe}^{-nt})$$

$$\dot{Q}^*(t) = (2b_2)^{-1}(2kb_1^2 + rb_1)X^*(t) + [b_1k + r - n]Q^*(t) - A(t), \qquad (2.5.17)$$

where $A(t) = (2b_2)^{-1}[a(2kb_1 - n + r) - (b_1k\mathbf{p} - b_1kc - rc)e^{-nt}]$. These equations may be interpreted as the optimal reaction curves of the dominant firm and the potential rivals in differential form. Note that in contrast with Gaskins' model, an optimal path for outputs can be derived from equations (2.5.17) (several cases are considered in the next section).

More insight about the long run properties of the model is provided by equations (2.5.17). At the steady state ($\dot{X}^* = \dot{Q}^* = 0$), the long run market share of the dominant firm is given by:

$$S = n(kb_1 - n + r)[rkb_1 - n^2 + nr + 2k(kb_1 - n + r)mv]^{-1}$$

which explains why the difference between the discount rate of the dominant firm (r) and the growth rate of the market (n) has also a crucial impact on the long run market share of the firms. It is easily shown that whenever $r < n$, a decrease in long run market share will result from an increase in risk aversion (m) or an increase in uncertainty (v) or an increase in the response coefficient (k) or a decrease in the growth rate (n).

These results are consistent with Sylos' and Gaskins' models. In contrast with earlier models, the price behavior is not always clearly established (Gaskins' model predicts a positive relationship between S and k; his assumption that the response coefficient takes the time-varying form $k = k_0 e^{nt}$ reduces the resulting system of differential equations corresponding to equations (2.5.17) to an autonomous system, which in turn allows specific conclusions about long run price behavior). It can also be proved that changes in the market share of the dominant firm is positively related to changes in the price elasticity of demand. This result is consistent with Pashingian's model which predicts that the rate of increase in the market share of the dominant firm is negatively related to changes in the estimated price elasticity of demand.

Simulation of Convergent Paths

The general solution of the system (2.5.17), when entrepreneurs are risk neutral is easily obtained. With $b = b_1 = b_2$, the system can be written:

$$\begin{bmatrix} 1 & 0 \\ 0 & 1 \end{bmatrix} \begin{pmatrix} \dot{X} \\ \dot{Z} \end{pmatrix} + \begin{bmatrix} \frac{1}{2}kb + n & \frac{1}{2}k^2 b \\ \frac{1}{2}b & -\frac{1}{2}kb + r - n \end{bmatrix} \begin{pmatrix} X \\ Y \end{pmatrix} = \begin{pmatrix} A_1 \\ A_2 \end{pmatrix}$$

where, $A_1 = ½ak + (c/2 - \mathbf{p})ke^{-nt}$, and $A_2 = ½(a - c)e^{-nt}$. The characteristic equation of the associated homogeneous system is given by $\det(M) = 0$, where

$$M = \begin{bmatrix} \frac{1}{2}kb + n - \theta & \frac{1}{2}k^2 b \\ \frac{1}{2}b & -\frac{1}{2}kb + r - n - \theta \end{bmatrix}.$$

The characteristic equation is; $\theta^2 + (2n - r)\theta + (-bkr/2 - nr + n^2) = 0$, which has roots:

$$\theta_1, \theta_2 = -n + r/2 \pm ½(\mathrm{r}2 + 2bkr)^{½}.$$

Assuming a solution of the form $X = Ce^{\theta t}$, $Z = KCe^{\theta t}$, a nontrivial particular solution must satisfy

$$M\begin{pmatrix}1\\K\end{pmatrix}C_1e^{\theta t}=0.$$

The constants K_1 and K_2, solutions of the above equations associated with the eigenvalues θ_1, θ_2, are: $K_1 = (bk + r + ½(r2 + 2bkr)^{½})/bk^2$, and $K_2 = (bk + r - ½(r2 + 2bkr)^{½})/bk^2$. The particular solutions of the homogeneous system are $X_1 = c_1e^{\theta 1t}$, $Z = K_1c_1e^{\theta 1t}$ and $X_2 = c_2e^{\theta 2t}$, $Z = K_2c_2e^{\theta 2t}$, and the general solution is:

$$X = C_1e^{\theta 1t} + C_2e^{\theta 2t} \qquad \text{and} \qquad Z = K_1C_1e^{\theta 1t} + K_2C_2e^{\theta 2t}$$

Solving for a particular solution of the non homogeneous system by the method of variation of constants;

$$C_1e^{\theta 1t} + C_2e^{\theta 2t} = A_1 \qquad \text{and} \qquad K_1C_1e^{\theta 1t} + K_2C_2e^{\theta 2t} = A_2\,.$$

Using Cramer's rule yields: $C_1' = (A_1K_2 - A_2)(K_2 - K_1)^{-1}e^{\theta 1t}$ and $C_2' = (A_1K_2 - A_2)(K_2 - K_1)^{-1}e^{\theta 2t}$. After substitution of A_1, A_2, K_1, K_2, the two constants can be integrated, which gives:

$$C_1 = ak(r - d)(2(r^2 + 2bkr + rd - 2nd)^{-1}e^{(2n - r - d)t/2}$$

$$+ ((c - 2\mathbf{p})(bk + r - d) + bkc)(2(r^2 + 2bkr + rd))^{-1}ke^{(d - r)t/2} + B_1$$

$$C_2 = ak(r + d)(2(r^2 + 2bkr - rd + 2nd)^{-1}e^{(2n - r + d)t/2}$$

$$+ ((c - 2\mathbf{p})(bk + r + d) + bkc)(2(r^2 + 2bkr - rd))^{-1}ke^{(d - r)t/2} + B_2$$

where, $d = (r^2 + 2bkr)^{½}$ and B_1, B_2 are constants of integration. Combining the general and particular solutions for X and Z and after simplification yields:

$$X^*(t) = B_1e^{½(r-2n+d)t} + B_2e^{½(r-2n-d)t} + e^{-nt}(bkc + rc - bk\mathbf{p} - 2r\mathbf{p})(rb)^{-1}$$

$$+ ak(r - n)(bkr + 2nr - 2n^2)^{-1} \tag{2.5.18}$$

$$Z^*(t) = (bk^2)^{-1}[(bk + r + (r^2 + 2bkr)^{½})B_1e^{½(r-2n+d)t} + (bk + r - (r^2 + 2bkr)^{½})]B_2e^{½(r-2n-d)t}$$

$$+ r^{-1}(c - \mathbf{p})e^{-nt} - (bkr + 2nr - 2n^2)^{-1}an \tag{2.5.19}$$

The constants B_1 and B_2 can be obtained from the initial conditions;

$$X(0) = B_1 + B_2 + (bkc + cr - pbk - 2pr)(br)^{-1} + ak(r - n)(bkr + 2nr - 2n^2)^{-1}$$

$$Z(0) = (bk + r + d)bk^2B_1 + (bk + r - d)(bk^2)^{-1}B_2 + (c - p)r^{-1} - an(bkn + 2nr - 2n^2)^{-1}\,.$$

Using Cramer's rule yields;

$$B_1 = (-2d)^{-1}[X(0) - (bkc + cr - \mathbf{p}bk - 2pr)(br)^{-1} - ak(r - n)(bkr + 2nr - 2n^2)^{-1}](bk + r - d)$$

$$- bk^2[Z(0) - (c - p)r^{-1} - an(bkn + 2nr - 2n^2)^{-1}]$$

$$B_2 = (-2d)^{-1}[X(0) - (bkc + cr - \mathbf{p}bk - 2pr)(br)^{-1} - ak(r - n)(bkr + 2nr - 2n^2)^{-1}](bk + r + d)$$

$$- bk^2[Z(0) - (c - p)r^{-1} + an(bkn + 2nr - 2n^2)^{-1}].$$

It can easily be shown that stability of the solution without oscillations is guaranteed by the condition $|2n - r| > |(r^2 + 2bkr)^{½}|$.

Four cases of simulated trajectory are presented below. In all these cases, the parameter values are so chosen as to satisfy the necessary condition of stability without oscillation. In all these cases, the marginal cost (c), initial output ($X(0)$), limit price (**p**) and static demand parameters (a and b) were kept fixed, and only the response coefficient (k), the discount rate (r) and market growth rate (n) were changed. In all cases, the initial output of rivals ($Z(0)$) was set at zero. The simulations reveal the following characteristics of the model:

* The time to converge to the steady state depends upon $X^*(0)$, $Z^*(0)$, and upon the extent by which the absolute value of $(2n - r)$ exceeds the absolute value of $(r^2 + 2bkr)^{½}$. An increase of either $Z^*(0)$ or of n will speed the rate of convergence to the steady state.

* The output and pricing policy is found to be very sensitive to variations in k (the strength of potential entry parameter) but not so much to the variations in the discount parameter (r) or market growth (n). As plots 1 and 3 illustrate, a change in the growth rate of demand and the discount rate does not change the price and output paths significantly when the response coefficient is large (k was set at .25 for these two simulations). As shown in the second simulation, a firm with a low response coefficient will rapidly loose market share and the market price will increase at nearly the same rate as the growth rate of demand. In such a case, the position of price leadership of the firm will quickly be challenged by rivals and the initial assumptions of the model might no longer hold.

In all simulations, the firm initially looses market share. In the long run, the firm can completely loose its market (simulation 2), maintain a fairly constant share (simulation 4), or slowly regain the initial loss after the entry of rivals (simulations 1 and 3).

Conclusions

In contrast with most existing limit price models, output rather than price is the major decision variable. In this model, price lies somewhere between the short-run monopoly price and the competitive price, the exact positioning depends on the barriers to entry, risk aversion and other behavioral attributes of the market. This model helps to explain why excessive profits can exist over time, but its predictions about specific dominant firm's behavior are not explicit. Unlike Gaskins' model, no definite relationship can be established between the quality of information about the demand parameters and price.

It was also found that an increase in risk aversion by the dominant firm leads to a lower short and long run output level, leads to a lower price in the long run, and accelerates the rate of convergence to the steady state. The simulated profiles show that in cases when there is a positive market growth, A variety of phases are possible in the optimal strategy space of the dominant firm which can in the long run increase, maintain or reduce its output.

TABLE 1: Trajectories for simulation 1.

Time	$e^{-nt}x^*(t)$	$e^{-nt}z^*(t)$	$e^{-nt}q^*(t)$	$p^*(t)$
0.0	1.0	5.1	140.0	5.51
10.0	10.3	68.2	144.0	5.36
20.0	15.9	111.0	149.0	5.34
30.0	19.3	158.0	134.0	5.41
40.0	21.5	204.0	129.0	5.50
50.0	23.0	249.0	123.0	5.62
60.0	24.2	282.0	117.0	5.75
70.0	25.1	335.0	112.0	5.89
80.0	26.0	377.0	106.0	6.03
90.0	26.7	418.0	101.0	6.17
100.0	27.4	458.0	95.6	6.31
110.0	28.1	498.0	90.3	6.45
120.0	28.8	538.0	85.1	6.59
130.0	29.4	576.0	79.9	6.72
140.0	30.0	614.0	74.8	6.85
150.0	30.6	652.0	69.8	6.99
160.0	31.2	689.0	64.9	7.11
170.0	31.8	725.0	60.1	7.24
180.0	32.4	761.0	55.3	7.37
190.0	33.0	797.0	50.6	7.49
200.0	33.6	831.0	45.9	7.61

Note: The parameters are: $x^*(0)=1.00$, $b=0.03$, $k=0.25$, $c=1.0$, $r=0.15$, $p=2.0$, $a=10.0$, $n=0.055$. Time step size is 10 units, number of steps is 21, and $z^*(t)$ is adjusted for sign.

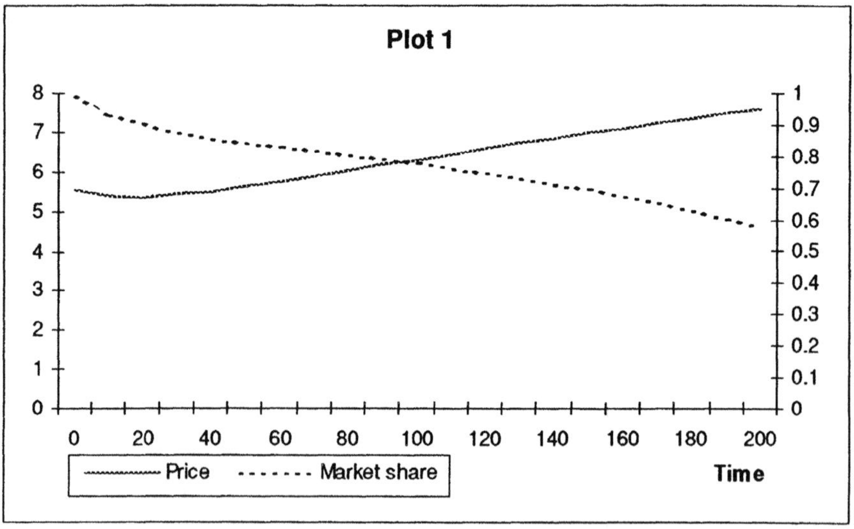

TABLE 2: Trajectories for simulation 2.

Time	$e^{-nt}x^*(t)$	$e^{-nt}z^*(t)$	$e^{-nt}q^*(t)$	$p^*(t)$
0.0	1.00	0.5	150.0	5.48
10.0	0.601	58.0	156.0	5.29
20.0	0.361	112.0	160.0	5.17
30.0	0.217	163.0	163.0	5.11
40.0	0.130	213.0	164.0	5.06
50.0	0.078	262.0	165.0	5.04
60.0	0.047	318.0	166.0	5.02
70.0	0.028	357.0	166.0	5.01
80.0	0.017	483.0	166.0	5.01
90.0	0.010	449.0	166.0	5.00
100.0	0.006	494.0	167.0	5.00
110.0	0.003	539.0	167.0	5.00
120.0	0.003	584.0	167.0	5.00
130.0	0.001	627.0	167.0	5.00
140.0	0.0008	617.0	167.0	5.00
150.0	0.0005	614.0	167.0	5.00
160.0	1.0003	757.0	167.0	5.00
170.0	0.0002	799.0	167.0	5.00
180.0	0.0002	841.0	167.0	5.00
190.0	0.0001	882.0	167.0	5.00
200.0	0.0001	923.0	167.0	5.00

Note: The parameters are: $x^*(0)=1.00$, $b=0.03$, $k=0.01$, $c=1.0$, $r=0.05$, $p=2.0$, $a=10.0$, $n=0.051$. Time step size is 10 units, number of steps is 21, and $z^*(t)$ is adjusted for sign.

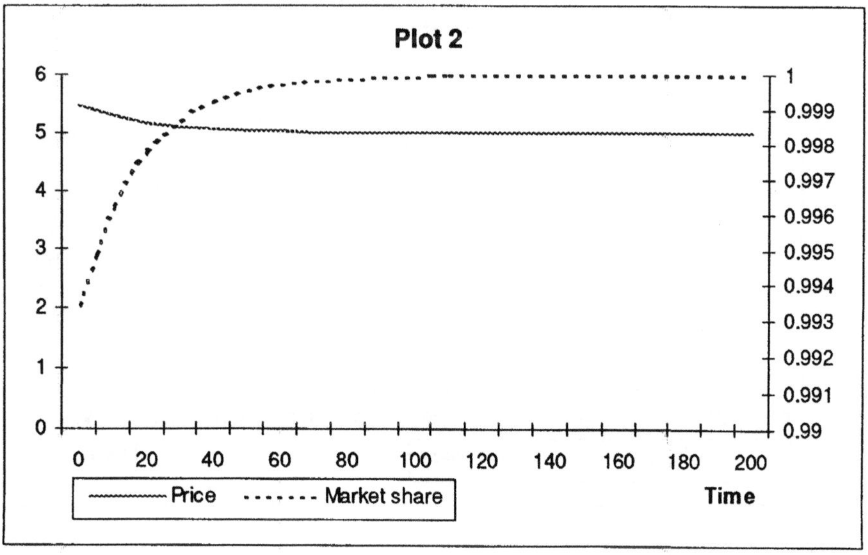

TABLE 3: Trajectories for simulation 3.

Time	$e^{-nt}x^*(t)$	$e^{-nt}z^*(t)$	$e^{-nt}q^*(t)$	$p^*(t)$
0.0	1.0	1.7	149.0	5.49
10.0	8.2	86.4	150.0	5.27
20.0	10.0	107.0	146.0	5.30
30.0	12.2	155.0	140.0	5.47
40.0	12.8	203.0	135.0	5.58
50.0	13.4	251.0	129.0	5.74
60.0	13.4	298.0	122.0	5.91
70.0	14.2	345.0	110.0	6.25
80.0	14.6	439.0	104.0	6.42
90.0	15.0	439.0	104.0	6.42
100.0	15.4	485.0	98.3	6.59
110.0	15.9	532.0	92.3	6.76
120.0	16.3	578.0	86.3	6.92
130.0	16.7	624.0	80.3	7.09
140.0	17.1	670.0	74.4	7.26
150.0	17.5	716.0	68.5	7.42
160.0	17.9	761.0	62.6	7.59
170.0	18.3	806.0	56.7	7.75
180.0	18.6	852.0	50.9	7.91
190.0	19.0	807.0	45.1	8.08
200.0	19.4	941.0	39.3	8.24

Note: The parameters are: $x^*(0)=1.00$, $b=0.03$, $k=0.25$, $c=1.0$, $r=0.10$, $p=2.0$, $a=10.0$, $n=0.104$. Time step size is 10 units, number of steps is 21, and $z^*(t)$ is adjusted for sign.

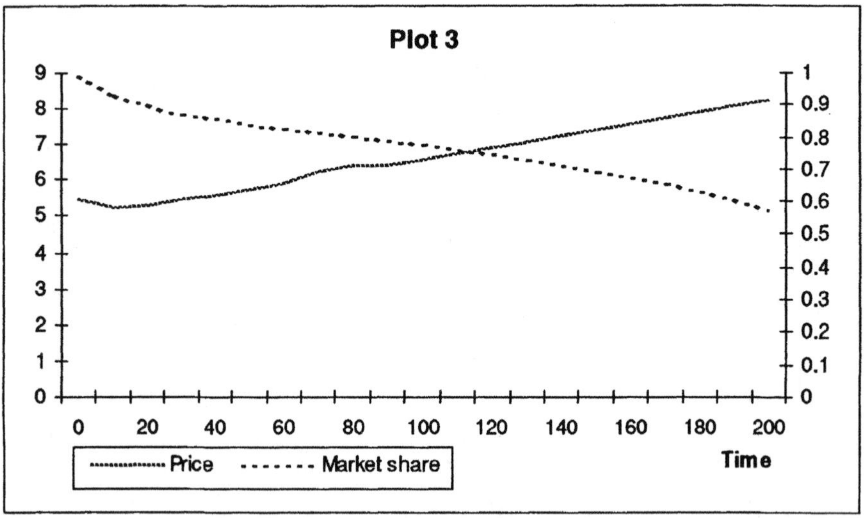

TABLE 4: Trajectories for simulation 4.

Time	$e^{-nt}x^*(t)$	$e^{-nt}z^*(t)$	$e^{-nt}q^*(t)$	$p^*(t)$
0.0	1.00	2.2	149.0	5.49
10.0	4.23	55.0	153.0	5.28
20.0	5.96	98.1	154.0	5.20
30.0	6.89	134.0	154.0	5.17
40.0	7.41	165.0	153.0	5.18
50.0	7.71	192.0	153.0	5.19
60.0	7.90	216.0	152.0	5.22
70.0	8.01	236.0	151.0	5.24
80.0	8.10	254.0	150.0	5.26
90.0	8.16	270.0	149.0	5.28
100.0	8.21	284.0	148.0	5.30
110.0	8.24	296.0	148.0	5.32
120.0	8.28	307.0	147.0	5.34
130.0	8.31	316.0	147.0	5.35
140.0	8.33	325.0	146.0	5.36
150.0	8.35	332.0	146.0	5.37
160.0	8.37	338.0	146.0	5.38
170.0	8.39	344.0	145.0	5.39
180.0	8.40	349.0	145.0	5.40
190.0	8.41	354.0	145.0	5.40
200.0	8.42	358.0	145.0	5.41

Note: The parameters are: $x^*(0)=1.00$, $b=0.03$, $k=0.10$, $c=1.0$, $r=0.05$, $p=2.0$, $a=10.0$, $n=0.064$. Time step size is 10 units, number of steps is 21, and $z^*(t)$ is adjusted for sign.

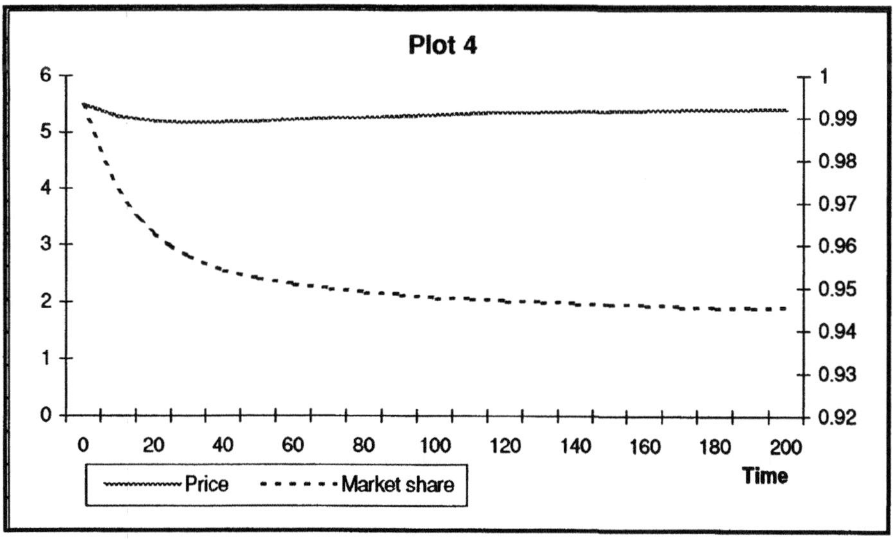

2.6 Example 2; reaching a steady state cycle

In general, the steady state is a single value identified with the help of the transversality conditions. However, using the transversality conditions can lead to faulty results when the control or the state variable are constrained. Following Fanchon and Melese (1996), we illustrate through a simulation how the steady state can be either a single value or a cycle.

The objective of a firm is to minimize labor costs when a fixed stock of in-house labor, $L_2(t)$, when more productive "outside" labor, $L_1(t)$, is available at a competitive market wage of w_1 per unit. The firm's decision to train its labor force is given by the control variable $u(t) \in [0,1]$, which can be interpreted as the intensity of in-house training (e.g. if u=1, then in-house labor is trained 100% of the time). The firm's output per unit of time is given by the additively separable production function,

$$Q(t) = sK + X\,L_1(t) + (1 - \beta\,u(t))x(t)\,L_2(t) \tag{2.6.1}$$

where: s is the productivity of capital, X is the productivity of contracted outside labor, $L_1(t)$; $x(t)$ is the productivity of in-house labor, L_2; and the stock of capital, K, is fixed. The parameter $0 \le \beta \le 1$ reflects the opportunity cost of training; If $\beta = 0$, then on-the-job training (OJT) does not interfere with the production process. Conversely, if $\beta = 1$, then when $u(t)$=1, L_2 does not contribute to production while it is being trained.

The OJT training decision, $u(t)$ has an impact on the productivity of in-house labor, $x(t)$, that enters the firm's production function, through the state equation:

$$dx/dt = u(t)\{1 - [x\,(t)/Y] + \delta x(t)\} - \delta x(t) \tag{2.6.2}$$

where the initial in-house labor productivity (prior to OJT) is given by $x(0)= x_0$, and the upper limit of productivity that can be achieved in the industry, Y, is greater than or equal to outside labor productivity, X; or $x_0 \le X \le Y$. (Note that equation (2.6.2) embodies diminishing returns to OJT investment). The last term on the right hand side represents human capital depreciation in the absence of training i.e. where: $\delta \ge 0$).

Once its in-house labor is trained, the firm might have to adjust in-house wages, $w_2(t)$, over time in order to retain its labor. The in-house wage adjustment is a function of in-house productivity, $x(t)$, relative to the productivity, X, of trained "outside" candidates, whose market wage is w_1. The in-house wage adjustment equation is thus given by:

$$w_2 = x_0(w_1/X) + (1 - \alpha)[x(t) - x_0](w_1/X), \tag{2.6.3}$$

where $\alpha \in [0,1]$ is the "loyalty" coefficient[16]. The first term on the right hand side of (3) is simply the starting wage of in-house workers, while the second term reflects productivity adjustments generated by OJT.

Assuming physical capital is fixed, the firm's objective is to minimize its discounted total labor costs of producing a given level of output. The firm's decision involves how intensively and how long to train its in-house labor (with wages given by equation (2.6.3)), when it has the alternative of hiring/retaining outside labor whose marginal productivity is X. The possible actions are to train its workers, and to hire, terminate, or retain outside labor.

OJT costs have two components; a fixed "setup" cost incurred every time training starts, and a variable cost that depends on the number of workers being trained. Let c denote the training cost per worker per unit of time, and d denote the occasional "setup" cost. The firm's labor costs are given by the sum of outside, in-house and OJT costs, or $C(t) = w_1L_1(t)+ w_2L_2(t) + cu(t)L_2(t)$. The first two terms reflect the wage costs of using outside and in-house labor respectively. The direct cost of OJT at time t is given by, $cu(t)L_2$, plus the setup cost. Given the desired production rate (Q), equation (2.6.1) can be used to express labor cost as:

$$C = (w_1/X)[Q - sK + \{\alpha x_0 - (\alpha - \beta u)x\}L_2] + cuL_2 \qquad (2.6.4)$$

The firm's problem can be stated as follows. Choose the amount of training $u(t)$ that minimizes the discounted present costs of producing Q per time period, subject to the change in in-house productivity described by equation (2.6.2).

The present value Hamiltonian is given by $H^d = H_0+H_1$, where $H_0 = (w_1/X)[Q\text{-}sK\text{-}\alpha(x\text{-}x_0)]\text{-}\lambda\delta x$, and $H_1 = [(w_1/X)\beta xL_2 + cL_2 + \lambda(1\text{-}Y^{-1}x+\delta x)]u$. H_0 represents the part over which the firm has no control (is independent of u), while H_1 represents the part that can be directly influenced by the actions of the firm (i.e. is a function of the control u). The first order conditions are:

$$dx/dt = \partial H^d/\partial t = u(1 - Y^{-1}x + \delta x) - \delta x \qquad (2.6.5)$$

and:

$$d\lambda/dt = r\lambda - \partial H^d/\partial x = [(Y^{-1} - \delta)u + r + \delta]\lambda + (w_1/X)L_2(\alpha - \beta u) \qquad (2.6.6)$$

where r is the discount rate.

Since the Hamiltonian is linear in the control, application of the maximum principle leads to a "bang-bang" solution where either u=0 (when H_1>0), or u=1

16 In the extreme case of "zero loyalty" (i.e. α=0), employers must adjust in-house wages to reflect all OJT-derived productivity gains to keep their workers. In the other extreme of "complete loyalty" (i.e. α=1), in-house workers exhibit complete wage flexibility, and employers can continue to offer the starting wage $w_2(0)$ regardless of any in-house productivity gains.

(when $H_1<0$). Hence only two options are available: either the entire in-house workforce is trained, or no one is trained.

The sign of H_1, depends upon the value of the costate variable λ, usually obtained by integrating equation (2.6.6) backward in time, starting from the steady state. Since the optimal control is "bang-bang", the control is either constant or alternates between two values (in which case the steady state is not uniquely defined and the costate equation is of no value since it cannot be integrated).

Assuming the system is time invariant, only three options are possible in the steady state: in-house labor is never trained (u=0), in-house labor is trained continuously (u=1), or a cycle of training and no-training is established (u alternates between 0 and 1). Each option is examined below.

Option 1: no training (u=0)
If in-house training stops as $x(t) = x_{00}$, equation (2.6.5) has the solution: $x(t) = x_{00}e^{-\delta t}$. Without training, in-house labor productivity decreases asymptotically to zero at the rate of depreciation of human capital δ, while the in-house wage converges to $X^{-1} w_1 x_0 \alpha$. The corresponding steady state labor cost per unit of time is given by:

$$C_0 = (w_1/X)(Q - sK) + (w_1/X)x_0\alpha L_2. \qquad (2.6.7)$$

The first term represents the cost of using outside labor to produce Q, while the second term is the fixed cost of paying non-productive in-house workers the minimum wage.

Option 2: constant training (u=1)
If in-house training stops as $x(t) = x_{01}$, equation (2.6.5) has the solution: $x(t) = (x_{01} - Y)e^{-t/Y} + Y$. Workers' productivity increases asymptotically to the upper limit of productivity, $x_{ss1} = Y$, (the solution of $dx/dt=0$). The corresponding cost per time unit is:

$$C_1 = C_0 + L_2[(c + \beta Y(w_1/X)) - \alpha Y(w_1/X)] \qquad (2.6.8)$$

The first term in brackets represents the direct cost and the opportunity costs of training a worker. The second term represents per-worker cost savings due to increased productivity. Equation (2.6.8) implies that the costs with constant training are lower than costs without training when $c < Y(w_1/X)(\alpha - \beta)$.

Option 3: periodic training
The first two options correspond to optimal values of the control variable, u=0 or u=1, leading to steady states asymptotically. An important third option is to use periodic training to maintain in-house workers' productivity within some bounds. In the steady state, firms alternate between periods of training (u=1) and no training

$(u=0)$[17]. In the training period, productivity is raised up to some optimal maximum level denoted by M. In the following period, productivity is allowed to decay to a minimum level denoted by m, at which point the cycle is repeated. The time path of productivity is shown in figure 8. The length of the non-training period, denoted by t_0, is given by: $t_0 = \delta^{-1} \ln (M/m)$, and the length of the training period, denoted by t_1, is similarly obtained: $t_1 = Y^{-1}\ln[(Y - m)/(Y - M)]$. The total length of the cycle is therefore $T = t_0 + t_1$.

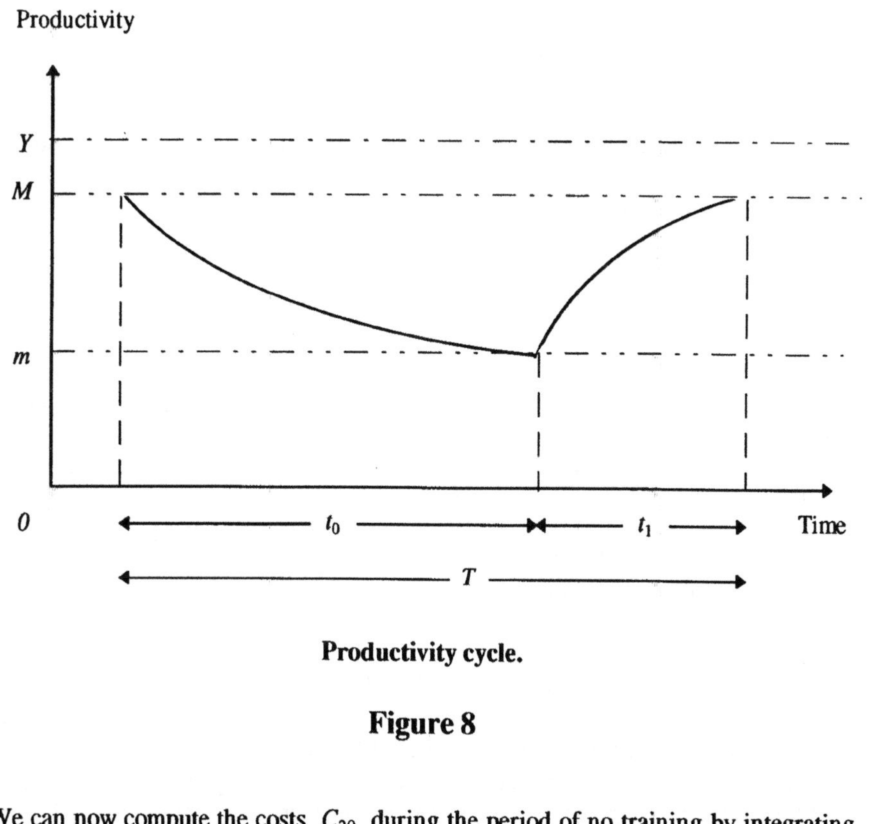

Productivity cycle.

Figure 8

We can now compute the costs C_{20} during the period of no training by integrating the cost function over the interval $[0,t_0]$. Similarly, the cost C_{21} during the period of training is equal to the setup cost, d, plus the integral of the cost function over the interval $[0,t_1]$. Finally, the average cost per unit of time over the entire cycle is given by $C_2 = T^{-1}(C_{20} + C_{21})$, which finally reads:

---

17 The steady state solution of option 3 satisfies the maximum principle without the transversality conditions. The solution is therefore "piecewise optimal" (as described in Seierstad and Sydsaeter [1987], p234).

$$C_2 = C_0 + \frac{1}{T}\left(d + cL_2 Y \ln\left[\frac{Y-m}{Y-M}\right]\right) + \beta Y L_2 \frac{w_1}{XT}\left\{Y \ln\left[\frac{Y-m}{Y-M}\right] - (M-m)\right\}$$
$$+\alpha L_2 \frac{w_1}{XT}\left\{\left(Y-\delta^{-1}\right)(M-m) - Y^2 \ln\left[\frac{Y-m}{Y-M}\right]\right\} \qquad (2.6.9)$$

where ln [.] denotes the natural logarithm. Labor costs per unit of time during the cycle can be separated into four components; the first term on the right hand side of equation (2.6.9) is the labor costs without training, the second term is the direct cost of training, the third term is the indirect cost of training, and the last term is the cost saving generated by worker "loyalty".

Note that since both the direct and indirect costs of training are positive, periodic training will be cost minimizing only if during the cycle, the cost saving due to worker loyalty is greater than the extra costs of training. The productivity triggers at which training starts m, and stops M, are determined by solving the set of first order conditions: $\partial C_2/\partial M = 0$ and $\partial C_2/\partial m = 0$.

The firm's optimal training decision is determined by choosing the option corresponding to the lowest of the steady state costs C_0, C_1, and C_2. It is difficult to obtain explicit values for M and m. However, under certain conditions it is possible to rank the costs of the different options. In particular, it is easy to prove that a cost minimizing firm will never provide training if its in-house workers have no loyalty (in that case periodic training always has higher steady state costs than not training in-house labor[18]). Finally, it can be established (using l'Hopital's rule) that C_2 converges to C_0 as M and m approach zero, that C_2 converges to C_1 as M and m approach Y, and that the length of the training cycle becomes infinite as m approaches zero, or as m approaches Y.

Comparative statics and simulation results:

Using (2.6.9), we can easily establish that an increase in capital reduces the costs of each option by the same amount ($\partial C_0/\partial K = \partial C_1/\partial K = \partial C_2/\partial K = - sw_1/X < 0$), and that $\partial C_2/\partial \beta > 0$, $\partial C_2/\partial c > 0$, which confirm the intuitive conclusion that an increase in either the direct or indirect costs of training also increase the cost of the periodic training option. However, the impact of a change in the loyalty coefficient α, or in the productivity of outside labor X, is uncertain. To obtain further insights into the behavior of the model, we ran a simulation with the

[18] Hence, if a firm does not have a loyal workforce (α=0), this model agrees with Becker's conclusion that that firms will not offer general training unless it is entirely financed by the workers. A corollary is that periodic training (option 3) will only be observed in firms or organizations where in-house labor demonstrates some loyalty.

following initial parameter values: $\alpha=.8$, $\beta=.1$, $c=.3$, $d=5$, $\delta=.05$, $K=100$, $L_2=500$, $Q=1000$, $s=1$, $w_1=1$, $X=.8$, $x_0=.7$, and $Y=1$. Using a grid, we generated the cost surface described by Equation (2.6.9). The surface shown in figure 9 represents the labor cost per unit of time using these initial values. We then changed the value of one of the parameters and observed how the surface evolved from its original shape, and how the corresponding values of m and M (trigger productivity values at which training starts and stops respectively) changed.

The comparative statics results obtained indicate that both the intensity and the duration of OJT are sensitive to various model parameters; including the mobility of in-house labor, the firm's discount rate, initial in-house labor productivity (relative to "outside" productivity), reduced productivity due to training, the wage for contracted labor, and direct training costs.

With small changes in the parameters, the simulation could generate all three possible outcomes; i.e. cases in which either no training, constant training, or cyclical training were the cost-minimizing solution. Cyclical training was the most common cost-minimizing option (figure 9), followed by the no training option (figure 10).

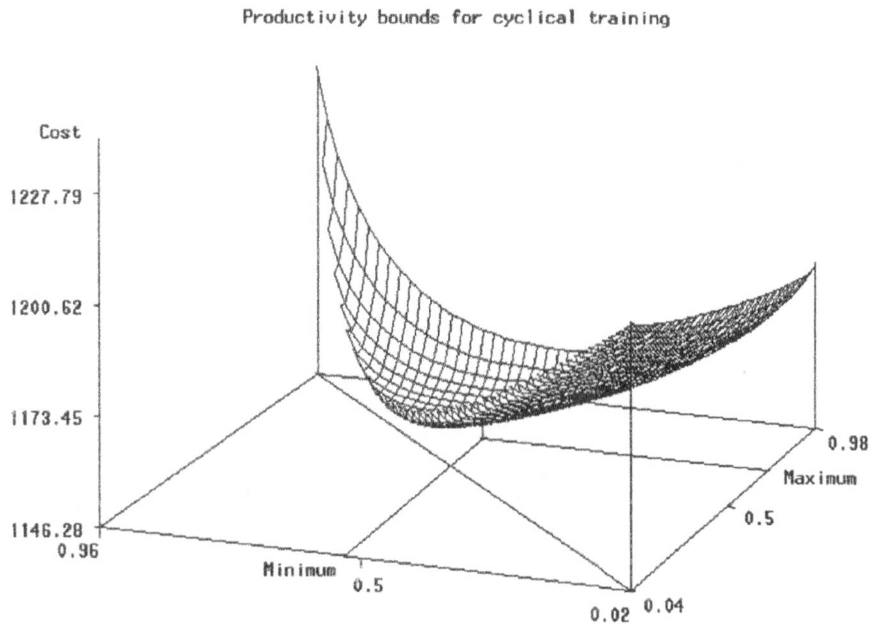

Case where cyclical training is optimal

Figure 9

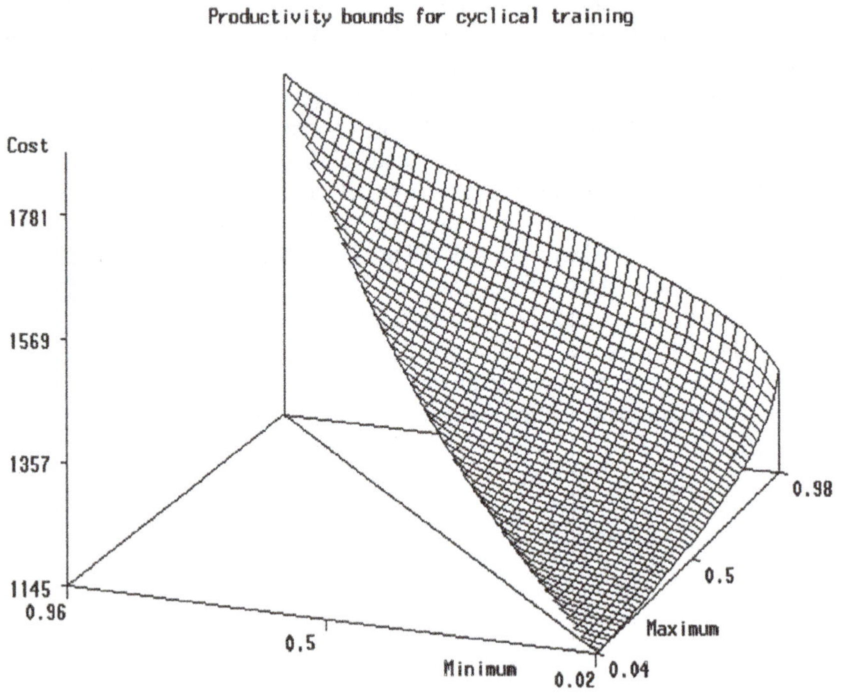

Case where no training is optimal

Figure 10

Constant training was preferred only in extreme cases where both the training and setup costs were zero or near zero. The no-training option was preferred when either the human capital depreciation rate was very high, or the loyalty coefficient was very low. The cost of the cyclical training option grows at a decreasing rate with an increase of the depreciation rate. In all the simulations, an increase in loyalty (α is increased from 0 to 1) leads to an increase in both the upper and lower productivity bounds of the training cycle (figure 11).

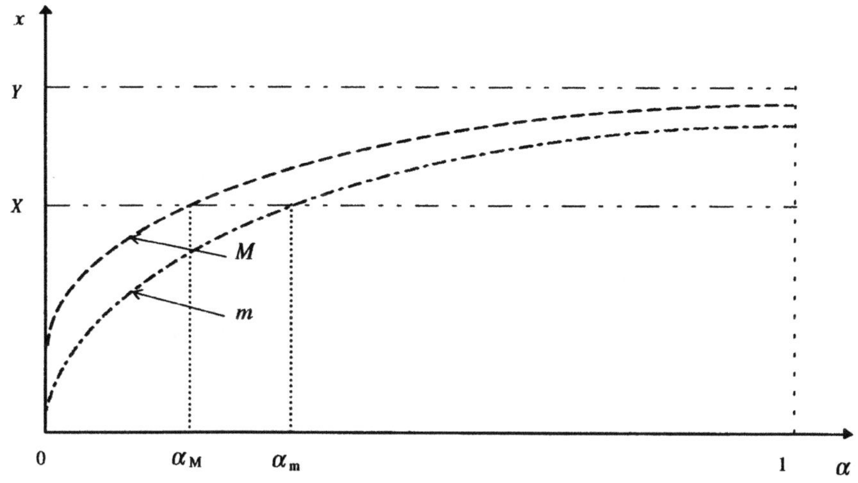

Upper and lower productivity bounds of the training cycle

Figure 11

The simulations also consistently show that with a sufficiently high loyalty coefficient, it is optimal to raise in-house worker productivity above that of outside workers ($X<M$, and in the extreme, $X<m<M$). The spread between the two bounds M and m remains fairly constant for the upper range of values of the loyalty parameter (for example from $\alpha=.5$ to $\alpha=1$). The results are very sensitive to the value of the decay factor (δ) and to the value of the loyalty coefficient (α). Periodic training outperforms pure strategies of either constant training or no training only if workers loyalty is high.

The optimal length of the training/non-training cycle is positively correlated with setup costs. The simulations reveal that providing OJT is not cost minimizing, unless the loyalty parameter exceeds a certain threshold. Hence it is possible that a firm will not offer any OJT even when it does not have to completely compensate workers for productivity gains.

References

Arrow, K. and Kurz, M, (1971), *Public investment, the rate of return, and Optimal Fiscal Policy.* Baltimore: John Hopkins Press.

Beltrami, E. (1987), *Mathematics for Dynamic Modeling.* Academic Press.

Burghes, D, and Graham, (1980), A. *Introduction to Control Theory including Optimal Control.* Ellis Horwood Series in Mathematics & its Applications, John Wiley and Sons.

Chiarella, Carl. (1990) *The elements of a nonlinear theory of economic dynamics.* Lecture Notes in Economics and Mathematical Systems vol. 343. Springer.

Dockner, E., Feichtinger, G. (1991) "On the Optimality of Limit Cycles in Dynamic Economic Systems". *Journal of Economics (Zeitschrift fur Nationalokonomie)*; 53(1), pages 31-50.

Fanchon, P. and Melese, F. (1996). "A Model of Periodic Job Training. " *International Advances in Economic Research*, vol 2.

Fanchon, P. Rifkin, E. and Sengupta, J. (1987). "A Dynamic and Stochastic Model of Price Leadership." *Developments of Control Theory for Economic Analysis*, Carraro and Sartore editors, Martinus Nijhoff (Kluwer), pages 239-260.

Goodwin, R. (1991). "*New Results in Non-linear Economic Dynamics*", Economic Systems Research; 3(4), pages 426-27.

Halkin, H. (1974), "Necessary Conditions for Optimal Problems with Infinite Horizons," *Econometrica*, vol. 42, pages 267-272.

Michel, P. (1982), "On the Transversality Condition in Infinite Horizon Optimal Problems", *Econometrica*, vol. 50, No. 5, pages 975-985.

Seierstad, A. And Sydsaeter, K., (1987), *Optimal Control Theory with Economic Applications*, North Holland.

Slotine, J. and Li, W., (1991) *Applied Non-Linear Control*, Prentice Hall.

Pontryagin, L. , Boltyanskii, V, Gramklelidze, R. and Mischenko, E. (1962), *The Mathematical Theory of Optimal Processes*, Wiley (Interscience) New York.

Wirl, F. (1994). "A New Route to Cyclical Strategies in Two-dimensional Optimal Control Models." *Ricerche Economiche*; 48(2), pages 165-73.

Wirl, F., (1994). "The Ramsey Model Revisited: The Optimality of Cyclical Consumption and Growth." *Journal of Economics (Zeitschrift fur Nationalokonomie)*; 60(1), pages 81-98.

CHAPTER 3: **Discrete time models**

3.0 Introduction

The control problems introduced in chapter 2 assume that a continuous flow of information is available to help us determine the best actions. With the assumption of continuity of the state and information vectors, we could use tools from the calculus and differential equations to identify the optimal solution. However, all economic data is collected at periodic intervals, and applications of the continuous control formulation in economics has been restricted almost exclusively to pure economic theory.

The discrete formulation introduced in this chapter is analogous to the continuous formulation of chapter 2, but uses information available at periodic points in time. As a consequence, the assumptions of piecewise continuity of the state and control variables made in the last chapter are no longer applicable. Initially, the problem seems far more complex than its continuous counterpart. Yet, with a small computer, the solution of a discrete control problem is far easier to identify, thanks to the vast body of linear algebra and econometric theory available. Most of all, because it is discrete, the theory introduced in this chapter can be used in numerous real life situations. In the discrete setting, the choice of control is made with the information available at time t, and the control cannot be changed until period t+1, when additional information becomes available. In the most general case, information about the state of the system is conveyed through the information vector: $y_t = Y(x_t)$. The information can be incomplete and imperfect for several reasons, for example because the system is stochastic, or some of its components are unobservable, or the information is noisy, or is incomplete. The uncertainty introduced by imperfections in the model or the information channels are usually modeled with the inclusion of an error term in the state and output equations. In sections 3.1 to 3.5, we temporarily assume that the information is perfect and complete, $y_t = x_t$, and that the system is deterministic. Section 3.6 concentrates on the problems associated with uncertain systems and "noisy" information channels. Section 3.7 presents several examples from economics, finance and business.

3.1 General deterministic and discrete control problems

With a discrete formulation of the control problem, the evolution of a system is now described by a set of difference equations:

$$x_{k+1} = f^k(x_k, u_k, z_k) \tag{3.1.1}$$

where x_k and u_k are the state and control vectors, and z_k is a vector of exogenous variables. The goal is to find an admissible control sequence $\{u_k\}_{k=1, 2, \ldots T}$ that would drive the system along a trajectory, minimizing an objective function chosen by the controller. Like the state equations (3.1.1), the objective function at time i, J_i, is also a discrete function of the state and control variables;

$$J_i = \phi(T, x_T) + \sum_{k=i}^{T-1} L^k(x_k, u_k) \tag{3.1.2}$$

Using an approach similar to the one used in the previous chapter, we form the Hamiltonian

$$H^k = L^k(x_k, u_k) + \lambda_{k+1} f^k(x_k, u_k). \tag{3.1.3}$$

and the solution of the following equations identify the optimal control sequence;

1/ state equation:

$$x_{k+1} = f^k(x_k, u_k, z_k)\,, \tag{3.1.4}$$

2/ costate equation:

$$\lambda_k = \frac{\partial H^k}{\partial x_k} = \frac{\partial L^k}{\partial x_k} + \left(\frac{\partial f^k}{\partial x_k}\right)' \lambda_{k+1}\,, \tag{3.1.5}$$

3/ Conditions for a minimum:

$$\underset{u_k \in U^k}{Min}\left(H^k(u_k)\right) \tag{3.1.6}$$

where U^k is the set of admissible controls at time k. If H^k is continuous in u and U^k is a simply connected admissible region, then condition (3.1.6) can be simplified to the stationarity condition:

$$\frac{\partial H^k}{\partial u_k} = \frac{\partial L^k}{\partial u_k} + \left(\frac{\partial f^k}{\partial u_k}\right)' \lambda_{k+1} = 0\,, \tag{3.1.6a}$$

Three of the equations above, ((3.1.4), (3.1.5) and either (3.1.6) or (3.1.6a)) are necessary to identify a minimum.

To completely identify the sequence of controls $\{u_k\}$, we also need two boundary conditions on the state or the costate vectors. The conditions can be imposed either as initial conditions or terminal conditions. The following two equations identify the relationships that must hold either at the start or at the end of the planning horizon.

a/ Initial conditions:

$$\left(\frac{\partial L^i}{\partial x_i} + \left(\frac{\partial f^i}{\partial x_i} \right)' \lambda_{i+1} \right)' dx_i = 0, \tag{3.1.7}$$

If the initial values of the state vector are known at the starting time i, then the system has a "fixed initial state". If x_i is fixed, then $dx_i = 0$, and (3.1.7) is always satisfied. This is the most common situation in control problems. When the initial conditions are not specified, the system has a "free initial state", and the expression multiplying dx_i must then be set equal to zero if the system is to satisfy the condition for a minimum at the initial time.

b/ Terminal conditions:

$$\left(\frac{\partial \phi}{\partial x_T} + \left(\frac{\partial \varphi}{\partial x_T} \right)' \mu_T - \lambda_T \right)' dx_T = 0. \tag{3.1.8}$$

A similar situation holds for the terminal conditions. The system has a *fixed final state* if the final values of the state vector must satisfy a pre-determined relationship $\varphi(x_T) = 0$. In such case, there are two constraints for the final period; the state vector must satisfy $\varphi(x_T) = 0$ and still satisfy the state equation (3.1.1). Hence the terminal conditions associated with the Hamiltonian, include two Lagrange multipliers; μ_T for the constraint $\varphi(x_T) = 0$, and λ_T for the constraint $x_T = f^T(x_{T-1}, u_{T-1})$. When the terminal constraint $\varphi(x_T) = 0$ is not specified, the system has a *free final state* and the Lagrange multiplier μ_T is equal to zero. However the state vector must still satisfy the state equation (3.1.1).

The terminal conditions for a control problem with an infinite horizon are called *transversality conditions* and are sufficient for the existence of a steady state, where all the components of the state vector become constant. Recent arguments have been made that the transversality conditions are not necessary for the existence of a solution, and empirical evidence shows that under certain conditions, a steady cycle might outperform a steady state.

3.1.1 The tracking problem.

An important class of control problem is the "tracking problem", frequently encountered in engineering, economics and finance. In a tracking problem, the decision maker wants to guide the system along a pre-determined path described by the target sequence $\{r_k\}$. The general solution of such non-linear problems is difficult to identify, and only limited qualitative results are available. However, some analytic results can be obtained in the special case when the target for each time period is a linear combination of some of the components of the state vector:

$$r_k = C_k x_k , \tag{3.1.9}$$

where r_k might involve only a subset of the components of x_k.

The objective function reflects the cost of not keeping the system on the desired path (i.e. the target sequence $\{r_k\}$). The original objective function (3.2.2) is therefore changed to:

$$J_i = \phi(T,(r_T - C_T x_T)) + \sum_{k=i}^{T-1} L^k((r_k - C_k x_k), u_k). \tag{3.1.10}$$

In this type of problem, there is no constraint that the state vector be a fixed value at any time (in other words, there is a cost for not meeting the target, but there is no obligation to meet the target).

In practical applications, the system to be controlled is usually non linear and sometimes unstable. The controller identifies a desired path using a stable model, and then designs a control law that will force the system along the path. The presence of unknown non linearities mandates a continuous update of the control law. Hence, in addition to defining a control law, the controller must also either identify an adaptation law for the control, or a revision of the model describing the system. The first option is called *adaptive control* design, and the second, *variable structure control* design. We provide a brief introduction to adaptive and variable structure control design in later chapters.

3.1.2 Controllability and observability.

Observability

Definition: The time invariant linear system (2.1.1) is *observable* if the initial state $x(0)$ can be uniquely determined from the output vectors $\{y(0), y(1), \dots y(k)\}$ for some finite integer k.

Controllability

Like observability, controllability is difficult to establish for non linear systems. It is not difficult to establish controllability for a linear system of the form (2.1.1). First, we must define the *(nxnm) state controllability matrix* as:

$$G = [B, AB, A^2 B, A^3 B, \dots A^{n-1} B]. \tag{2.1.3}$$

Controllability of the system (2.2.1) is guaranteed if and only if the rank of the state controllability matrix (denoted by $\rho[G]$) is less than the dimension of the state vector. Note however that it might take a long time for the state vector to reach its target value. With the additional constraint that the target must be reached within P periods, where P < n, the state controllability matrix becomes:

$$G_P = [B, AB, A^2B, A^3B, \dots A^{P-1}B], \tag{2.1.4}$$

and the system is controllable if and only $\rho[G_P] \leq P$. In both cases, there are no constraints on the possible values of the control vector.

Example 1

The state controllability matrix for a system of the form (3.2.1), where

$$A = \begin{bmatrix} 1 & 0 & 1 \\ 0 & 1 & 1 \\ 1 & 1 & 0 \end{bmatrix} \text{ and } B = \begin{bmatrix} 0 \\ 0 \\ 1 \end{bmatrix} \text{ is: } G = \begin{bmatrix} \begin{pmatrix} 0 \\ 0 \\ 1 \end{pmatrix} & \begin{pmatrix} 1 \\ 1 \\ 0 \end{pmatrix} & \begin{pmatrix} 0 \\ 1 \\ 2 \end{pmatrix} & \begin{pmatrix} 2 \\ 3 \\ 2 \end{pmatrix} & \cdots \end{bmatrix}.$$

The state vector has dimension equal to 3. Since $\rho[B] = 1$, $\rho[B, AB] = 2$, $\rho[B, AB, A^2B] = 3$, the system is state controllable within 3 periods. In other words, it will take at most three periods before the state vector assumes a predetermined value.

If in the previous system, the matrix B was instead $B = \begin{bmatrix} 0 & 1 \\ 0 & 0 \\ 1 & 0 \end{bmatrix}$, the state controllability matrix would become: $G = \begin{bmatrix} \begin{pmatrix} 0 & 1 \\ 0 & 0 \\ 1 & 0 \end{pmatrix} & \begin{pmatrix} 1 & 1 \\ 1 & 0 \\ 0 & 1 \end{pmatrix} & \begin{pmatrix} 0 & 1 \\ 1 & 1 \\ 2 & 1 \end{pmatrix} & \cdots \end{bmatrix}$.

Since $\rho[B] = 2$ and $\rho[B, AB] = 3$, the system is state controllable within 2 periods.

Example 2

Consider the system of example 1 where the matrix B is instead $B = \begin{bmatrix} 1 & -1 \\ -1 & 1 \\ 0 & 0 \end{bmatrix}$.

The state controllability matrix is now: $G = \left[\begin{pmatrix} 1 & -1 \\ -1 & 1 \\ 0 & 0 \end{pmatrix} \begin{pmatrix} 1 & -1 \\ -1 & 1 \\ 0 & 0 \end{pmatrix} \begin{pmatrix} 1 & -1 \\ -1 & 1 \\ 0 & 0 \end{pmatrix} \cdots \right]$.

Since $\rho[B] = \rho[B, AB] = \rho[B, A^2B] \ldots = 1$, the system is state uncontrollable.

For large systems, the dimension of the state controllability matrix can become enormous, and state controllability can be difficult to establish. A convenient theorem somewhat simplifies the task:

Theorem (Hautus, 1969):
The linear system (2.1.1) is state controllable if and only if the matrix $[(A - \lambda I)B]$ has rank n for every eigenvalue λ of A.

3.2 The linear quadratic problem

Like their continuous counterparts, analytic solutions of the general discrete optimal control problem are difficult to obtain. The control vector is usually expressed as a complicated function of the state vector, and a policy corresponding to such a control vector would be difficult to implement. If we impose the constraint that the system be linear, and use a quadratic approximation of the objective function, the resulting linear-quadratic (LQ) control problem has a simple solution. Practically, excellent results have been obtained using the LQ control as an approximation of non-linear systems. In some cases the symmetry of the quadratic objective function is inappropriate because the cost of exceeding a target is quite different from the cost of being short of the target. Such cases can be handled using an asymmetric objective function introduced by Friedman(1972).

Even with their two characteristic restrictions, linear quadratic control problems can cover a multitude of situations. Yet all the variations of LQ problems always have two things in common: the controlled system has linear dynamics, described by[19]:

$$x_{k+1} = A_k x_k + B_k u_k \qquad k = 1,2, \ldots T\text{-}1, \qquad (3.2.1)$$

and the objective function for period i is quadratic:

19 Since the most general linear models the form $x_{k+1} = A_k x_k + \ldots A_{k\text{-}p} x_{k\text{-}p} + B_k u_k + \ldots B_{k\text{-}q} u_{k\text{-}q} + z_k$ can easily be transformed into an equation with functional form (3.2.1), we only consider models of the form (3.2.1). Various methods available to reduce other formulations to (3.2.1) are presented in section 3.4.

$$J_i = \frac{1}{2}\left[x_T' S_T x_T + \sum_{k=i}^{T-1} \left(x_k' Q_k x_k + u_k R_k u_k \right) \right], \tag{3.2.2}$$

where $x_k \in R^n$, $u_k \in R^m$, A_k, B_k, Q_k, R_k, S_T are matrices of appropriate dimensions, and Q_k, R_k, S_T are symmetric positive semi-definite.

The most important variations are: the terminal time T can be either fixed or free (possibly infinite), the terminal conditions for x can be fixed or free, and constraints can be imposed on the possible values the control vector can take.

Before we provide a method of solution for the LQ problem, it is very instructive to determine, for a reference, what would be the behavior of the system if no control is applied. It is helpful to think of the objective function as a function measuring the "cost-to-go", that is the cumulative cost that would be incurred until the last period.

If no control is applied, the cost-to-go during the last period is: $J_T = \frac{1}{2} x_T' S_T x_T$. Using (3.2.1), the cost-to-go for period T-1 is: $J_{T-1} = \frac{1}{2} x_{T-1}' \left(A_{T-1}' S_T A_{T-1} + Q_{T-1} \right) S_T x_{T-1}$, which can also be written as $J_{T-1} = \frac{1}{2} x_{T-1}' S_{T-1} x_{T-1}$, where $S_{T-1} = A_{T-1}' S_T A_{T-1} + Q_{T-1}$. Repeating the process, we have the cost-to-go for period k:

$$J_k = \frac{1}{2} x_k' S_k x_k , \tag{3.2.3}$$

where

$$S_k = A_k' S_{k+1} A_k + Q_k . \tag{3.2.4}$$

The preceding recurrence for S_k can be integrated backwards from the terminal time to compute the sequence of matrices $\{S_k\}$. To do that, we need to know what the final value of the state vector would be if no control is ever applied. That value is readily available if we use (3.2.1) with $u_k \equiv 0$. If the state vector is x_i at the start of period i, at the terminal time T it will be:

$$x_T = \left(\prod_{j=i}^{T} A_j \right) x_i . \tag{3.2.5}$$

The cost-to-go for period k can then be computed backwards from the terminal state, using the recurrence relations (3.2.3) and (3.2.4).

In case the planning horizon is infinite, the final state x_T will be finite only if all the matrices A_k are stable for k greater than some finite number s. In other words, the final state x_T will be finite only if the eigenvalues of A_k have modulus less than 1 for all $k \geq s$. Consequently, as time becomes infinite, the vector x_i

reaches the steady state $x_{ss}=0$. If the system is time invariant, the matrix S_k converges to the steady state value $S_\infty = \sum_{i=0}^{\infty} (A')^i Q A^i$ with the corresponding steady state cost-to-go $J_\infty = \frac{1}{2} x_0' S_\infty x_0$.

It is quite interesting to note that starting from any point x, equations (3.2.3) to (3.2.5) are sufficient to compute ahead of time what will be the cost-to-go if the system is left to evolve by itself, without any control.

We now solve the LQ control problem. The Hamiltonian is

$$H^k = \frac{1}{2}\left(x_k' Q_k x_k + u_k R_k u_k\right) + \lambda_{k+1}'\left(A_k x_k + B_k u_k\right), \tag{3.2.6}$$

and we can use the first order conditions (3.1.3) to (3.1.8) to identify the optimal control sequence. The state and costate equations are

$$\frac{\partial H_k}{\partial \lambda_k} = x_{k+1} = A_k x_k + B_k u_k \tag{3.2.7}$$

$$\frac{\partial H_k}{\partial x_k} = \lambda_k = Q_k x_k + A_k' \lambda_{k+1} \tag{3.2.8}$$

and the stationary condition (which assumes that H_k is differentiable in u_k) is

$$\frac{\partial H_k}{\partial u_k} = 0 = R_k u_k + B_k' \lambda_{k+1}. \tag{3.2.9}$$

The last equation gives us a value for the control vector:

$$u_k = R_k^{-1} B_k' \lambda_{k+1} \tag{3.2.10}$$

which allows us to eliminate u_k in the other two equations. After a few manipulations[20], the system (3.2.7) and (3.2.8) can be written as:

$$\begin{bmatrix} x_k \\ \lambda_k \end{bmatrix} = \begin{bmatrix} A_k^{-1} & A_k^{-1} B_k R_k^{-1} B_k' \\ Q_k A_k^{-1} & A_k' + Q_k A_k^{-1} B_k R_k^{-1} B_k' \end{bmatrix} \begin{bmatrix} x_{k+1} \\ \lambda_{k+1} \end{bmatrix} \tag{3.2.11}$$

---

20 Use (3.2.10) in the state equation and express x_k as a function of x_{k+1} and λ_{k+1}, then use the resulting expression for x_k in (3.2.8).

provided A_k is non singular. The coefficient matrix on the right hand side is called the *Hamiltonian matrix*. Note that equation (3.2.11) has been developed independently of the boundary conditions, and therefore must hold no matter what the boundary conditions.

If we knew the values for x and λ at the terminal time, we could use the backward recurrence relation above to identify the paths of the state and costate variables, and ultimately the optimal control sequence using (3.2.9). The boundary conditions allows us to solve the system by providing the missing element.

The boundary conditions critically depend upon what we assume will prevail at the end of the planning horizon. If the value of x is not specified at the terminal time T (which is possibly infinite), then the linear-quadratic control problem is a free final state problem. Otherwise, it is a fixed final state problem. Finally, if T is infinite, the control problem has infinite horizon. We will now review each case in detail.

3.2.1 Case 1: fixed final state.

In this case, we have two constraints for the terminal conditions; the system must satisfy the state equation, and must also satisfy the terminal condition $\varphi(T,x_T) = 0$. Consequently, the Lagrange multiplier μ_T of the terminal conditions (3.1.8) is non-zero. In the linear-quadratic formulation, we impose the terminal condition that a that linear combination of some components of the state vector assume a given value at the terminal time:

$$Cx_T - r_T = 0 \tag{3.2.17}$$

From (3.1.8), the terminal condition is:

$$\lambda_T = S_T x_T + C'\mu_T \tag{3.2.18}$$

The recursion (3.2.11) is going backwards for the state and costate vectors. To solve it, we need to know the final values of the state and costate vectors. Unfortunately, that information is not available, and we must make some more assumptions to find a solution. An elegant method to solve the problem, the *sweep method* was proposed by Bryson and Ho. The method is based on the assumption that a linear relationship between the state and costate vectors exists at all times, and not just at the terminal time. That is we assume that (3.2.18) also holds for $k \le T$:

$$\lambda_k = S_k x_k + V_k \mu_T \tag{3.2.19}$$

where S_k and V_k are matrices to be determined, and μ_T is the (scalar) Lagrange multiplier of the terminal condition, also to be determined. Using (3.2.19), and after a few manipulations[21], we have:

$$S_k = A_k' S_{k+1}(A_k - B_k K_k) + Q_k \tag{3.2.20}$$

$$V_k = (A_k - B_k K_k)' V_{k+1}, \tag{3.2.21}$$

where the matrix K_k, called the *Kalman gain matrix* of the system, is defined as

$$K_k = (B_k' S_{k+1} B_k + R_k)^{-1} B_k' S_{k+1} A_k . \tag{3.2.22}$$

Since at the terminal time, both (3.2.19) and (3.2.18) are satisfied, it is clear that

$$V_T = C' . \tag{3.2.23}$$

Now that we have computed the matrices S_k and V_k introduced in (3.2.19), we must still determine the value of the (scalar) Lagrange multiplier μ_T. To that end, and like in the previous paragraph, we make the assumption that the terminal condition can also be expressed with the same functional form as (3.2.18):

$$r_T = U_k x_k + P_k \mu_T \tag{3.2.24}$$

for $k \le T$, where the matrices U_k and P_k are to be determined. Again, after more manipulations[22], U_k and P_k are given by

$$U_k = V'_k, \tag{3.2.25}$$

---

21 Use (3.2.19) in the state equation (3.2.7) and solve for x_{k+1},. Then use the expression for x_{k+1} and equation (3.2.19) in the costate equation and factor the terms in x_k and μ_T. Finally, use the matrix inversion lemma to simplify the expression multiplying x_k. Since the relationship must hold for any value of x_k , the coefficients of x_k and μ_T must both be equal to zero, hence equations (3.2.20) and (3.2.21).

22 Use (3.2.19) in the state equation (3.2.7) and solve for x_{k+1},. Since μ_T- is a constant, first differencing (3.2.19) yields $U_{k+1}x_{k+1} + P_{k+1}\mu_T - U_k x_k - P_k \mu_T = 0$. Substitute for x_{k+1}, , factor x_k. and μ_T-, and use the matrix inversion lemma to simplify the expression multiplying x_k. Since the relationship must hold for any value of x_k , the coefficients of x_k and μ_T must both be equal to zero. The terminal condition (3.2.17) implies equations (3.2.25) and (3.2.26).

and

$$P_k = P_{k+1} - V'_{k+1}B_k\left(B'_kS_{k+1}B_k + R_k\right)^{-1}B'_kV_{k+1} \tag{3.2.26}$$

with terminal conditions $V_T = C' = \mu_T$, and P_T=0. The optimal LQ control problem has a solution if all the sequence of matrices P_k are non singular, from the time of origin to the terminal time. After computing before hand the sequences of matrices $\{S_k\}$, $\{V_k\}$, $\{P_k\}$, and $\{U_k\}$, and after more rearrangements and substitutions in (3.2.10), we finally have an expression for the optimal control:

$$u_k = -\left(B'_kS_{k+1}B_k + R_k\right)^{-1}B'_k\left(V_{k+1}P_k^{-1}r_T + \left(S_{k+1}A_k - V_{k+1}P_k^{-1}V'_k\right)x_k\right). \tag{3.2.27}$$

Note that the control vector is determined by the current value of the state vector. Control laws of this form, called *closed loop control* laws, can take into account unforeseen fluctuations in the state vector. In contrast, a control sequence entirely determined at the initial time is called an *open loop control* law. Because open loop control laws cannot adapt to unforeseen changes in the state vector, their performance tend be much worse than closed loop control laws.

Example: We consider a time varying problem with three periods (T=3), where the matrices A_k, B_k, Q_k, R_k and S_k defined in (3.2.1) and (3.2.2) are:

$$A_1 = \begin{pmatrix} 1 & -.2 \\ .1 & .5 \end{pmatrix}, \quad A_2 = \begin{pmatrix} 1 & -.1 \\ 0 & .5 \end{pmatrix}, \quad B_1 = \begin{pmatrix} 2 & 2 \\ 2 & 1 \end{pmatrix}, \quad B_2 = \begin{pmatrix} 3 & 1 \\ 1 & 1 \end{pmatrix}$$

$$Q_1 = \begin{pmatrix} .1 & 0 \\ 0 & .1 \end{pmatrix}, \quad Q_2 = \begin{pmatrix} .3 & .1 \\ .1 & .2 \end{pmatrix}, \quad R_1 = \begin{pmatrix} .1 & .1 \\ .1 & .2 \end{pmatrix}, \quad R_2 = \begin{pmatrix} .2 & 0 \\ 0 & .1 \end{pmatrix} \quad S_3 = \begin{pmatrix} 1 & 0 \\ 0 & .5 \end{pmatrix}.$$

The initial conditions are $x_1 = \begin{pmatrix} 1 \\ 1 \end{pmatrix}$, and the terminal conditions are r_T=Cx_T, where

$$C = \begin{pmatrix} 1 & -1 \\ 1 & 1 \end{pmatrix} \quad \text{and} \quad r_T = \begin{pmatrix} 1 \\ 3 \end{pmatrix}.$$

Before computing the solution, we must verify that the matrices Q_1, Q_2, R_1, R_2, and S_3 are positive definite. The eigenvalues of Q_1, Q_2, R_1, R_2, and S_3 are respectively (.10,.10), (.36,.14), (.26,.04), (.20,.10), and (1,.5), which ensures that the conditions are met.

Next, we compute the matrices K_2, S_2, V_2, K_1, S_1, V_1, using (3.2.20) to (3.2.22). Starting the recursion with S_3, and V_3, which are known (recall that $V_3=C'$ from the terminal conditions), we have:

$$K_2 = \begin{pmatrix} .40 & -.19 \\ -.24 & .50 \end{pmatrix}, \quad S_2 = \begin{pmatrix} .35 & .06 \\ .06 & .25 \end{pmatrix}, \quad V_2 = \begin{pmatrix} .20 & -.10 \\ -.23 & .14 \end{pmatrix}, \quad P_2 = \begin{pmatrix} -1.96 & .37 \\ .37 & -2.67 \end{pmatrix},$$

$$K_1 = \begin{pmatrix} -.06 & .35 \\ .45 & -.38 \end{pmatrix}, \quad S_1 = \begin{pmatrix} .16 & -.04 \\ -.04 & .13 \end{pmatrix}, \quad V_1 = \begin{pmatrix} .10 & -.06 \\ -.07 & .04 \end{pmatrix}, \quad P_1 = \begin{pmatrix} -2.17 & .49 \\ .49 & -2.64 \end{pmatrix}.$$

Using (3.2.26), we compute the optimal control and state vectors:

$$u_1 = \begin{pmatrix} -.27 \\ -.07 \end{pmatrix}, \quad x_2 = \begin{pmatrix} .13 \\ 0 \end{pmatrix} \quad u_2 = \begin{pmatrix} .43 \\ .57 \end{pmatrix} \quad x_3 = \begin{pmatrix} 2 \\ 1 \end{pmatrix},$$

and the cost associated with the control sequence is J=4.79. Note that it is not possible to compare the cost with what it would have been if no control was applied since without control, the terminal condition would not be met.

3.2.2 *Case 2: free final state.*

The terminal conditions for the general case are given by (3.1.8). With no constraints on the final value of the state vector, we cannot garrote that the system will be in a steady state at the terminal time. Hence dx_T is normally non-zero in (3.1.8). Also, since we do not have the second constraints in the final state, the last period Lagrange multiplier, μ_T, of (3.1.8) is equal to zero and the terminal conditions are satisfied if $\lambda_T = \dfrac{\partial \phi}{\partial x_T}$. In the linear case, the condition reduces to:

$$\lambda_T = S_T x_T \,. \tag{3.2.28}$$

The system defined by (3.2.11) and (3.2.28) is a boundary value problem for which we know the initial and final values of the state vector. Like in the previous section, we use the "sweep method" to identify the solution. Hence, we assume that a linear relationship between the state and costate vectors exists at all times, and not just at the terminal time. That is we assume that

$$\lambda_k = S_k x_k \tag{3.2.29}$$

also holds at all times prior to T. Using (3.2.27) and (3.2.10) in the state equation leads to a forward recursion for the state vector. With our assumption of a linear relationship between the state and costate vectors, the forward recursion leads to a backward recursion for the matrices S_k defined in (3.2.28):

$$S_k = A_k'\left[S_{k+1} - S_{k+1}B_k\left(B_k'S_{k+1}B_k + R_k\right)^{-1}B_k'S_{k+1}\right]A_k + Q_k. \tag{3.2.30}$$

The equation above, called the *Riccati equation*, allows us to compute the sequence of matrices $\{S_k\}$ from the terminal value S_T (which is specified in the objective function and therefore is known). Note that since we know all the matrices A_k, B_k, R_k and Q_k, it is possible to compute the sequence $\{S_k\}$ ahead of time. The sequence of optimal control vectors $\{u_k\}$ is obtained by solving the state equation (3.2.1) for u_k after substituting for the value of λ_k and S_k given in (3.2.10) and (3.2.30). After a few manipulations, the optimal control is given by:

$$u_k = -K_k x_k, \tag{3.2.31}$$

where:

$$K_k = \left(B_k'S_{k+1}B_k + R_k\right)^{-1}B_k'S_{k+1}A_k \tag{3.2.32}$$

is the Kalman gain matrix. Finally, the state trajectory is obtained by substituting the values of the optimal control vectors into the state equation. In this case, it is always instructive to compute the zero input cost, which can be used to assess the benefits of control.

The preceding results can easily be extended to the case where the objective function is homogeneous of degree two in the state and control vectors, i.e. when J_i is given by:

$$J_i = \frac{1}{2}\left[x_T'S_Tx_T + \sum_{k=i}^{T-1}\left(x_k'Q_kx_k + u_k'R_ku_k + u_k'V_kx_k + x_k'W_ku_k\right)\right]. \tag{3.2.33}$$

After the change of variables $T = ½(V'+W)$, the Riccati equation (3.2.30) and the Kalman gain matrix (3.2.32) corresponding to the new objective function respectively become:

$$S_k = \overline{A'}_k\left[S_{k+1} - S_{k+1}B_k\left(B_k'S_{k+1}B_k + R_k\right)^{-1}B_k'S_{k+1}\right]\overline{A}_k + \overline{Q}_k \tag{3.2.34}$$

and

$$K_k = \left(B_k'S_{k+1}B_k + R_k\right)^{-1}\left(B_k'S_{k+1}A_k + T_k'\right), \tag{3.2.35}$$

where

$$\overline{A}_k = A_k - B_kR_k^{-1}T_k' \tag{3.2.36}$$

and

$$\overline{Q}_k = Q_k - T_k R_k^{-1} T_k' . \tag{3.2.37}$$

Example: We consider the same time varying system of example 1, but without the terminal constraint. After verification that the matrices Q_1, Q_2, R_1, R_2, and S_3 are positive definite, we compute the matrices S_2 and S_1 from the Riccati recursion (3.2.4). Starting the recursion with S_3, which is known, we have:

$$S_2 = \begin{pmatrix} 56.3 & 7 \\ 7 & 2.51 \end{pmatrix} \quad \text{and} \quad S_1 = \begin{pmatrix} 48.61 & 13.68 \\ 13.68 & 6.36 \end{pmatrix}.$$

Using (3.2.16) , we can now compute the Kalman gain matrices:

$$K_1 = \begin{pmatrix} -.34 & .55 \\ .83 & -.65 \end{pmatrix} \quad \text{and} \quad K_2 = \begin{pmatrix} .40 & -.18 \\ -.24 & .50 \end{pmatrix}.$$

Equation (3.2.15) gives us the optimal controls: $u_1 = \begin{pmatrix} -.215 \\ -.183 \end{pmatrix}$ and $u_2 = \begin{pmatrix} -.004 \\ .008 \end{pmatrix}$.

Equation (3.2.1) gives us the state vectors: $x_1 = \begin{pmatrix} 1 \\ 1 \end{pmatrix}$, $x_2 = \begin{pmatrix} .002 \\ -.014 \end{pmatrix}$ and

$$x_3 = \begin{pmatrix} .001 \\ -.003 \end{pmatrix}.$$

Finally, the associated cost $J = .219$ is computed with (3.2.2). We also check the zero input cost (J_0) using (3.2.2): $J_0 = .560$. Hence applying the control sequence reduced the cost by .341 (or 60%).

3.2.3 *Case 3: The tracking problem*

In a tracking problem, the decision maker wants to guide the system along a pre determined path described by the target sequence $\{r_k\}$ where the target for each time period is a linear combination of some of the components of the state vector:

$$r_k = C_k x_k . \tag{3.2.38}$$

The objective function reflects the cost of not keeping the system on the desired path (i.e. the target sequence $\{r_k\}$). The original objective function (3.2.2) is therefore changed to

$$J_i = \frac{1}{2}\left[(C_T x_T - r_T)' P_T (C_T x_T - r_T) + \sum_{k=i}^{T-1}\left((C_k x_k - r_k)' Q_k (C_k x_k - r_k) + u_k R_k u_k\right)\right], \qquad (3.2.39)$$

where P and Q are positive semi-definite matrices, and R is positive definite. In this type of problem, there is no constraint that the state vector be a fixed value at any time (in other words, there is a cost for not meeting the target, but there is no obligation to meet the target).

Using (3.2.1) and (3.2.34), we form the Hamiltonian, which yields the first order conditions (3.2.1), (3.2.9) and

$$\frac{\partial H_k}{\partial x_k} = \lambda_k = C_k' Q_k C_k x_k - C_k' Q_k r_k + A_k' \lambda_{k+1}, \qquad (3.2.40)$$

and the terminal conditions are given by:

$$\frac{\partial H_T}{\partial x_T} = \lambda_T = C_T' P_T (C_T x_T - r_T). \qquad (3.2.41)$$

Like in the previous cases, we use the sweep method to identify a solution: Extending the functional relationship between the state and costate variables at the terminal time to all periods, we assume that

$$\lambda_k = S_k x_k - v_k \qquad (3.2.42)$$

where at the terminal time, we have:

$$S_T = C'_T P_T C_T \qquad \text{and} \qquad v_T = C'_T P_k r_T . \qquad (3.2.43)$$

After a few manipulations[23], we can identify S_k and v_k as:

$$S_k = A_k'\left[S_{k+1} - S_{k+1} B_k (B_k' S_{k+1} B_k + R_k)^{-1} B' S_{k+1}\right] A_k + C_k' Q_k C_k \qquad (3.2.46)$$

---

23 Substitute the value of λ_k in the state equation and solve for x_{k+1}. Then substitute λ_k and x_{k+1}. in the costate equation and factor x_k. Since the resulting expression must hold for all time periods, the coefficient of x_k and the constant must always be equal to zero, which yields expressions for S_k and v_k. Finally, apply the matrix inversion lemma gives (3.2.44) and (3.2.45).

$$v_k = \left[A_k' - A_k' S_{k+1} B_k \left(B_k' S_{k+1} B_k + R_k\right)^{-1} B'\right] v_{k+1} + C_k' Q_k r_k \tag{3.2.47}$$

We can now use the recurrences (3.2.41) and (3.2.42) to obtain the sequences $\{S_k\}$ and $\{v_k\}$. In the next step, we use (3.2.39) to compute the costate variables, and after a few more manipulations, we finally have the optimal control

$$u_k = -K_k x_k + K_{k+1}^a r_k \tag{3.2.48}$$

where the matrices K_k (the Kalman gain matrix) and K_k^a are:

$$K_k = \left[\left(B_k' S_{k+1} B_k + R_k\right)^{-1} B' S_{k+1}\right] A_k \tag{3.2.49}$$

and

$$K_k^a = \left(B_k' S_{k+1} B_k + R_k\right)^{-1} B'. \tag{3.2.50}$$

Starting with the initial value for the state, x_0, we compute u_0, then x_1, and so on.

Example: Again, we consider the time varying system of numerical example 1, but with the objective to track a path identified by equation (3.2.33), where

$$C_1 = \begin{pmatrix} 1 & -1 \\ 1 & 0 \end{pmatrix},\ C_2 = \begin{pmatrix} 1 & -1 \\ 0 & 1 \end{pmatrix},\ C_3 = \begin{pmatrix} 1 & -1 \\ 1 & 1 \end{pmatrix},\ r_1=(1,1)',\ r_2=(1,2)' \text{ and } r_3=(1,3)'.$$

After verification that the matrices Q_1, Q_2, R_1, R_2, and P_3 are positive definite, we compute S_3 and v_3 with the terminal condition (3.2.38):

$$S_3 = \begin{pmatrix} 1.5 & -.5 \\ -.5 & 1.5 \end{pmatrix} \quad \text{and} \quad v_3 = \begin{pmatrix} 2.5 \\ .5 \end{pmatrix}.$$

We then compute S_2, v_2, S_1 and v_1 from the Riccati recursions (3.2.39) and (3.2.40):

$$S_2 = \begin{pmatrix} .37 & -.26 \\ -.26 & .37 \end{pmatrix},\ v_2 = \begin{pmatrix} .53 \\ .01 \end{pmatrix},\ S_1 = \begin{pmatrix} .28 & -.15 \\ -.15 & .14 \end{pmatrix} \text{ and } v_1 = \begin{pmatrix} .31 \\ -.15 \end{pmatrix}.$$

Next, we compute the gain matrices with (3.2.42) and (3.2.43);

$$K_2^r = \begin{pmatrix} .24 & -.20 \\ .02 & .85 \end{pmatrix}, \quad K_2 = \begin{pmatrix} .46 & -.25 \\ -.40 & .67 \end{pmatrix}, \quad K_1^r = \begin{pmatrix} 1.23 & 3.07 \\ 1.09 & -1.30 \end{pmatrix}, \quad K_1 = \begin{pmatrix} -.26 & .48 \\ .66 & -.53 \end{pmatrix}.$$

Finally compute the controls and the resulting state vectors; $x_1=(1,1)'$, $u_1=(.46,.43)'$, $x_2=(2.58,1.95)'$, $u_2=(-.19,.18)'$, $x_3=(2.01,.97)'$. The tracking errors are $e_1=(1,0)'$, $e_2=(.37,.05)'$, $e_3=(-.04,.02)'$, and the associated cost is $J_1=.255$.

3.2.4 Case 4: The infinite horizon (steady state) problem

The solution of the infinite horizon problem parallels that of the fixed time free final state case presented earlier, with the difference that the final time is now infinite. With an infinite horizon, the system can either have an explosive behavior, be in a perpetual cycle or in a steady state as time becomes infinite[24]. To have a steady state, the system must be time invariant as t becomes infinite. Hence the Riccati equation (3.2.30) now reads:

$$S = A'\left[S - SB\left(B'SB + R\right)^{-1} B'S\right]A + Q, \tag{3.2.51}$$

Of course, there is no guarantee that a solution of the Riccati equation exists. If a solution S_{ss} does exist, it can be computed either by brute force (i.e. by iteration of (3.2.20), using an arbitrary initial value for the matrix S_k.) or with the eigenvalue method (presented in the next section). Some more elegant methods to solve the Riccati equation are available, but given the great speed of most personal computers, the two methods mentioned above work well for systems of small dimension.

As time becomes infinite, equations (3.2.31) and (3.2.32) become:

$$K_{ss} = \left(B'S_{ss}B + R\right)^{-1} B'S_{ss}A \tag{3.2.52}$$

$$u_{ss} = -K_{ss}x_{ss} \ , \tag{3.2.53}$$

where K_{ss} denotes the steady state Kalman gain, u_{ss} and x_{ss} are the steady state control and state vectors. Finally, the path to the steady state is given by:

$$x_{k+1} = A_{cl}x_k \tag{3.2.54}$$

where

---

24 Several conditions guaranteeing asymptotic stability of the system must be met. In particular, the system must be "reachable" and "stabilizable". A detailed definition of these concepts is outside the scope of this book and can be found in standard control books (e.g. F. L. Lewis, or C.T. Chen).

$$A_{cl} = \left(A - BK_{ss}\right). \tag{3.2.55}$$

The system will be asymptotically stable if the eigenvalues of A_{cl} have modulus less than unity. The system will be cyclical with periodicity m if A_{cl} is idempotent of order m, and the system is unstable if some of the eigenvalues of A_{cl} have modulus greater than unity.

3.3 Analytic solution of the Riccati equation

The numerous methods available to solve the Riccati equation can be divided in two large groups; iterative methods, and eigenvalue methods. The first group uses variations of Gauss-Seidel or gradient methods with an arbitrary initial value for the Riccati matrix. The methods are usually very computationally expensive, and are being replaced by the more efficient methods of the second group. The essence of eigenvalue methods is to use the eigenvalues and eigenvectors of the Hamiltonian matrix (defined in (3.2.11)), to identify the gain matrices without solving the Riccati equation. One of the first eigenvalue method widely accepted was proposed by MacFarlane [1963]. The method has since been refined by numerous authors (e.g. Vaughan [1970], Laub [1979], and Wang [1989]). A substantial survey of current algorithms can be found in Pierce [1986].

All the eigenvalue methods are based upon the fact that the Hamiltonian matrix is *simplectic*. By definition, a simplectic matrix satisfies $H'JH=J$, where $J = \begin{bmatrix} 0 & I \\ -I & 0 \end{bmatrix}$. Consequently, any simplectic matrix H has the property that $H^{-1}=-JH'J$, and the eigenvalues of H are also eigenvalues of H^{-1}. Hence if μ is an eigenvalue of H, then so is $1/\mu$. Because the Hamiltonian matrix is simplectic, its eigenvalues can be grouped in two sets; the first one containing the eigenvalues μ_1, μ_2, ... μ_n outside the unit circle, and the other one containing the corresponding eigenvalues $1/\mu_1$, $1/\mu_2$, ... $1/\mu_n$ inside the unit circle. Hence, H can be written as:

$$H = WDW^{-1}, \tag{3.3.1}$$

where

$$D = \begin{bmatrix} D_1 & 0 \\ 0 & D_1^{-1} \end{bmatrix}, \tag{3.3.2}$$

and

$$D_1 = \mathrm{Diag}(\mu_1, \mu_2, \dots \mu_n). \tag{3.3.3}$$

The matrix W is the matrix of corresponding eigenvectors, partitioned comformably with D as:

$$W = \begin{bmatrix} W_{11} & W_{12} \\ W_{21} & W_{22} \end{bmatrix}. \tag{3.3.4}$$

We can now use the decomposition of H to make the change of variable:

$$(y_k, \eta_k)' = W^{-1}(x_k, \lambda_k)'. \tag{3.3.5}$$

In the new coordinate system, the Hamiltonian equation (3.2.11) becomes

$$(y_k, \eta_k)' = D\,(y_{k+1}, \eta_{k+1})', \tag{3.3.6}$$

and the terminal conditions are: $x_T = W_{11}y_T + W_{12}\eta_T$ and $\lambda_T = W_{21}y_T + W_{22}\eta_T$. At the terminal time, we must have:

$$\lambda_T = S_T x_T. \tag{3.3.7}$$

After substitution for λ_T and x_T, and after solving for η_T, the relationship above reads:

$$\eta_T = N y_T, \tag{3.3.8}$$

where

$$N = -(W_{22} - S_T W_{12})^{-1}(W_{21} - S_T W_{11}). \tag{3.3.9}$$

We can now extend the relationship (3.3.8) to the non terminal periods, and identify the matrices N_k satisfying a relationship of the form $\eta_k = N_k y_k$. If the system is time varying, the eigenvalues and eigenvectors of the Hamiltonian matrix must be computed for every period, a very tedious process which greatly complicates the solution. Hence, for the sake of clarity, we only present the time invariant case, and the reader can refer to more advanced control textbooks for a treatment of some time varying cases. Starting from the terminal time, (3.3.6) yields:

$$y_T = D_1^{-(T-k)} y_k \tag{3.3.10}$$

and

$$\eta_k = D_1^{-(T-k)} \eta_T \tag{3.3.11}$$

Substitution of (3.3.8) and (3.3.10) in (3.3.11) leads to the desired relationship, where

$$N_k = -D_1^{-(T-k)}(W_{22} - S_T W_{12})^{-1}(W_{21} - S_T W_{11})D_1^{-(T-k)}. \tag{3.3.12}$$

Finally, we can express the relation $\eta_k = N_k y_k$ in terms of the original variables x_k and λ_k, using the inverse transformation of (3.3.5). The resulting coefficient matrix

$$S_k = (W_{21} + W_{22}T_k)(W_{11} + W_{12}N_k)^{-1} \tag{3.3.13}$$

relating the state and costate variables ($\lambda_k = S_k x_k$) is the solution S_k of the Riccati equation. Note that to compute the matrices S_k, we only need the eigenvalues and eigenvectors of the (time invariant) Hamiltonian matrix. Yet, the computational burden can still be enormous when dealing with large systems common in economics and finance (especially in the time varying case). The quest for a simple efficient algorithm is still alive although all the proposed improvements are well beyond the scope of this book[25].

In the infinite horizon case, a necessary condition for the steady state to exist is that the constraint be non binding in the steady state. In other words, the costate variables must be zero as time becomes infinite. This will only happen if the matrix D_1^{-1} of (3.3.10) is stable (i.e. D_1 is unstable). Since D_1 has all the eigenvalues of H outside the unit circle on its diagonal, the condition is clearly met, and in the limit, D_1^{-k} becomes the null matrix as k becomes infinite. The matrix N_k of (3.3.12) also becomes the null matrix, and the solution of the Riccati equation given by (3.3.13) degenerates to

$$S_\infty = W_{21} W_{11}^{-1} \tag{3.3.14}$$

which is the solution of the Algebraic Riccati Equation, computed only from the unstable eigenvectors of the Hamiltonian matrix.

Example: We consider a time invariant problem, where the matrices A, B, Q, R and S_T defined in (3.2.1) and (3.2.2) are:

$$A = \begin{pmatrix} 1 & 1 \\ .284 & 2.644 \end{pmatrix}, \quad B = \begin{pmatrix} 0 \\ .284 \end{pmatrix}, \quad Q = \begin{pmatrix} 1 & 0 \\ 0 & .2 \end{pmatrix}, \quad R = .5, \quad S_T = \begin{pmatrix} 1 & 0 \\ 0 & 1 \end{pmatrix}.$$

The initial conditions are $x_0 = (.2, .1)'$, and the terminal conditions are free. From (3.2.11), the Hamiltonian matrix is:

[25] The solution proposed by Wang (1989) is accessible to a reader with good command of intermediate linear algebra.

$$H = \begin{pmatrix} .903 & -.342 & 0 & -.055 \\ .097 & .342 & 0 & .055 \\ .903 & -.342 & 1 & -.339 \\ .019 & .068 & 1 & 2.655 \end{pmatrix}.$$

The eigenvalues are: $(\mu_1, \mu_2, \mu_3, \mu_4) = (2.375, 1.378, .421, .726)$, and corresponding eigenvectors are:

$$v_1 = \begin{pmatrix} -.042 \\ .024 \\ -.270 \\ .962 \end{pmatrix}, \quad v_2 = \begin{pmatrix} .089 \\ -.024 \\ .784 \\ -.614 \end{pmatrix}, \quad v_3 = \begin{pmatrix} .549 \\ .754 \\ -.339 \\ .124 \end{pmatrix}, \quad v_4 = \begin{pmatrix} -.451 \\ -.170 \\ .782 \\ -.395 \end{pmatrix}.$$

The matrices D and W of (3.3.2) and (3.3.4) are:

$$D = \begin{pmatrix} 2.375 & 0 & 0 & 0 \\ 0 & 1.378 & 0 & 0 \\ 0 & 0 & .421 & 0 \\ 0 & 0 & 0 & .726 \end{pmatrix} \quad \text{and} \quad W = \begin{pmatrix} -.042 & .089 & .549 & -.451 \\ .024 & -.024 & .754 & -.170 \\ -.270 & .784 & -.339 & .782 \\ .962 & -.614 & .124 & -.395 \end{pmatrix}.$$

The solution of the infinite horizon Riccati equation is given by (3.3.14);

$$S_\infty = W_{21} W_{11}^{-1} = \begin{pmatrix} -.270 & .784 \\ .962 & -.614 \end{pmatrix} \begin{pmatrix} -.042 & .089 \\ .024 & -.024 \end{pmatrix}^{-1} = \begin{pmatrix} 10.936 & 7.888 \\ 7.404 & 53.041 \end{pmatrix}.$$

3.4 Equivalent Dynamic Equations

The dynamics of many systems are described by an equation different from

$$x_{k+1} = A_k x_k + B_k u_k . \tag{3.4.1}$$

which is the one used throughout most of this book. The different formulations are usually not a problem since most of them can readily be reduced to (3.4.1).

3.4.1 *High order linear systems*

With a simple two step process, a linear system with a high order dynamic equation of the form:

$$x_{k+1} = A_k x_k + A_{k-1} x_{k-1} + \cdots + A_{k-m} x_{k-m} + B_k u_k + B_{k-1} u_{k-1} + \cdots + B_{k-n} u_{k-n}. \quad (3.4.2)$$

can be reduced to an augmented first order system with dynamic equation of the form (3.4.1). First, augment the dynamic equation with m-1 identities $x_i = x_i$ written as:

$$\begin{array}{lclllllll} x_k & = & Ix_k + Ox_{k-1} & + \cdots + Ox_{k-m+1} + Ox_{k-m} & + Ou_k + Ou_{k-1} + \cdots + Ou_{k-n} \\ x_{k-1} & = & Ox_k + Ix_{k-1} & + \cdots + Ox_{k-m+1} + Ox_{k-m} & + Ou_k + Ou_{k-1} + \cdots + Ou_{k-n} \\ \vdots & & \vdots \quad \vdots & \quad \vdots \quad \vdots & \vdots \quad \vdots \quad \vdots \\ x_{k-m+1} & = & Ox_k + Ox_{k-1} & + \cdots + Ix_{k-m+1} + Ox_{k-m} & + Ou_k + Ou_{k-1} + \cdots + Ou_{k-n} \end{array}$$

where I and O are identity matrices and null matrices of appropriate dimensions. Then rewrite the system in terms of the augmented state and control vectors z_k and w_k:

$$z_{k+1} = \Theta_k z_k + \Omega_k w_k \quad (3.4.3)$$

where

$$z_k = (x_k', x_{k-1}', x_{k-2}', \cdots, x_{k-m}')' \quad (3.4.4)$$

$$w_k = (u_k', u_{k-1}', u_{k-2}', \cdots, u_{k-n}')' \quad (3.4.5)$$

$$\Theta_k = \begin{bmatrix} A_k & A_{k-1} & \cdots & A_{k-m} \\ I & O & \cdots & O \\ \vdots & \ddots & \ddots & \vdots \\ O & O & I & O \end{bmatrix} \quad (3.4.6)$$

$$\Omega_k = \begin{bmatrix} B_k & B_{k-1} & \cdots & B_{k-n} \\ O & O & \cdots & O \\ \vdots & \vdots & \ddots & \vdots \\ O & O & O & O \end{bmatrix} \quad (3.4.7)$$

The dynamic equation of the augmented system (3.4.3) is now identical to (3.4.1).

Example: Let the following equation describe the dynamics of a system

$$\begin{pmatrix} x_1 \\ x_2 \\ x_3 \end{pmatrix}_{k+1} = \begin{bmatrix} 1 & 5 & 0 \\ 1 & 3 & 1 \\ 0 & 0 & 2 \end{bmatrix} \begin{pmatrix} x_1 \\ x_2 \\ x_3 \end{pmatrix}_k + \begin{bmatrix} 2 & 2 & 3 \\ 0 & 2 & 1 \\ 2 & 0 & 1 \end{bmatrix} \begin{pmatrix} x_1 \\ x_2 \\ x_3 \end{pmatrix}_{k-1} + \begin{bmatrix} 2 & 0 \\ 2 & 1 \end{bmatrix} \begin{pmatrix} u_1 \\ u_2 \end{pmatrix}_k . \tag{3.4.8}$$

From (3.4.4) and (3.4.5), we have $z_k = (x_{1k}, x_{2k}, x_{3k}, x_{1k-1}, x_{2k-1}, x_{3k-1})'$, $w_k = u_k$, and (3.4.8) can be written as:

$$\begin{pmatrix} x_{1,k+1} \\ x_{2,k+1} \\ x_{3,k+1} \\ x_{1,k} \\ x_{2,k} \\ x_{3,k} \end{pmatrix} = \begin{bmatrix} 1 & 5 & 0 & 2 & 2 & 3 \\ 1 & 3 & 1 & 0 & 2 & 1 \\ 0 & 0 & 2 & 2 & 0 & 1 \\ 1 & 0 & 0 & 0 & 0 & 0 \\ 0 & 1 & 0 & 0 & 0 & 0 \\ 0 & 0 & 1 & 0 & 0 & 0 \end{bmatrix} \begin{pmatrix} x_{1,k} \\ x_{2,k} \\ x_{3,k} \\ x_{1,k-1} \\ x_{2,k-1} \\ x_{3,k-1} \end{pmatrix} + \begin{bmatrix} 2 & 0 \\ 2 & 1 \\ 0 & 0 \\ 0 & 0 \\ 0 & 0 \\ 0 & 0 \end{bmatrix} \begin{pmatrix} u_{1,k} \\ u_{2,k} \end{pmatrix}, \tag{3.4.9}$$

and the state equation now has the same functional form as (3.4.1).

3.4.2 System with exogenous variables

The dynamic equation of many models in economics, finance and operations research include an exogenous term:

$$x_{k+1} = A_k x_k + B_k u_k + c_k. \tag{3.4.10}$$

Like in the previous paragraph, the dynamic equation can also be reduced to the form (3.4.1) by augmenting the state vector and using an identity. Define

$$\chi_k = \begin{pmatrix} x_k \\ 1 \end{pmatrix}, \quad \mathcal{A}_k = \begin{bmatrix} A_k & c_k \\ \phi & 1 \end{bmatrix} \quad \text{and} \quad \Xi_k = \begin{bmatrix} B_k \\ \varphi \end{bmatrix}, \tag{3.4.11}$$

where ϕ and φ are null row vectors of appropriate dimension. The augmented system now reads:

$$\chi_{k+1} = \mathcal{A}_k \chi_k + \Xi_k u_k , \tag{3.4.12}$$

which is of the desired form. The last row of (3.4.12) is simply the identity $\{1=1\}$, and the remaining rows are the initial system (3.4.10).

Example: Let the following equation describe the dynamics of a time invariant system:

$$x_{k+1} = \begin{pmatrix} x_1 \\ x_2 \end{pmatrix}_{k+1} = \begin{bmatrix} 2 & 3 \\ 4 & 5 \end{bmatrix} \begin{pmatrix} x_1 \\ x_2 \end{pmatrix}_k + \begin{bmatrix} 1 & 1 \\ 0 & 2 \end{bmatrix} \begin{pmatrix} u_1 \\ u_2 \end{pmatrix}_k + \begin{pmatrix} 2 \\ 5 \end{pmatrix}. \qquad (3.4.13)$$

Using the transformation describe above, the system becomes:

$$\chi_{k+1} = \begin{pmatrix} x_1 \\ x_2 \\ 1 \end{pmatrix}_{k+1} = \begin{bmatrix} 2 & 3 & 2 \\ 4 & 5 & 5 \\ 0 & 0 & 1 \end{bmatrix} \begin{pmatrix} x_1 \\ x_2 \\ 1 \end{pmatrix}_{k+1} + \begin{bmatrix} 1 & 1 \\ 0 & 2 \\ 0 & 0 \end{bmatrix} \begin{pmatrix} u_1 \\ u_2 \end{pmatrix}_k, \qquad (3.4.14)$$

which now has the desired functional form.

3.4.3 *System design with a prescribed degree of stability*

One of the fundamental properties of linear quadratic systems under closed loop optimal control is their stability. Even though such systems are always stable, their performance can degrade severely with a poor estimate of the parameters, with time delays, with unmodelled linearities and with random perturbations. Such is the case when the eigenvalues of the closed loop state feedback matrix A_{cl} have negative real parts close to zero. In order to guarantee that the system will remain stable, the policy maker might want to use a control law which guarantees that the corresponding eigenvalues of the matrix A_{cl} are all less than a specified negative quantity θ. It is easy to prove[26] that the closed loop state feedback matrix $A_{\theta cl}$ of the system with the modified state equation

$$x_{k+1} = (A+\theta I)x_k + Bu_k \qquad (3.4.15)$$

has such property (I denotes the n x n identity matrix). The optimal control $u_{\theta k}$ minimizing the original objective function is given by:

$$u_{\theta k} = R^{-1}B'P_\theta x_k = -K_\theta x_k, \qquad (3.4.16)$$

where P_θ is the positive semidefinite solution of the algebraic Riccati equation in which the matrix A has been replaced by $(A+\theta I)$. Since all the eigenvalues of the corresponding closed loop state feedback matrix $A_{\theta cl} = (A+\theta I)-BK_\theta$ have negative real parts, it follows that all the eigenvalues of the original closed loop state feedback matrix $A_{cl} = A-BK_\theta$ also have negative real parts smaller than $-\theta$. Hence the system under the control law (3.4.10) will have the desired level of stability.

---

[26] For example, Patel and Mundo (1982).

Example: Let the equation

$$\begin{pmatrix} x_1 \\ x_2 \end{pmatrix}_{k+1} = \begin{bmatrix} 0 & 1 \\ -.284 & 1.644 \end{bmatrix} \begin{pmatrix} x_1 \\ x_2 \end{pmatrix}_k + \begin{bmatrix} 0 \\ .284 \end{bmatrix} u_k \qquad (3.4.17)$$

describe the dynamics of a system with the quadratic objective function $J = \Sigma_{k=1,2,}(x_k'Qx_k+u_k'Ru_k)$. If a degree of relative stability of $\theta = -.5$ is sought, equation (3.4.15) becomes

$$\begin{pmatrix} x_1 \\ x_2 \end{pmatrix}_{k+1} = \begin{bmatrix} .5 & 1 \\ -.284 & 2.144 \end{bmatrix} \begin{pmatrix} x_1 \\ x_2 \end{pmatrix}_k + \begin{bmatrix} 0 \\ .284 \end{bmatrix} u_k \qquad (3.4.18)$$

The solution P_θ of (3.4.17) and the associated gain K_θ are:

$$P_\theta = \begin{bmatrix} 86.031 & 26.697 \\ 26.697 & 52.631 \end{bmatrix} \quad \text{and} \quad K_\theta = [9.478 \quad 18.685]. \qquad (3.4.19)$$

We can now use K_θ and the original matrix A to obtain the closed loop state feedback matrix $A_{cl} = A\text{-}BK_\theta$ which has all its eigenvalues with negative real parts smaller than -.5 (the eigenvalues of A_{cl} are -2.445 and -1.216).

3.5 Discrete control of nonlinear systems

The vast majority of systems are nonlinear. When compared to linear systems, nonlinear systems are generally more difficult to control because models cannot capture all the nonlinear dynamics. Only in a few cases (mostly in physics and engineering) can some laws perfectly describe the system to be controlled. This is why modellers and controllers often approximate nonlinear systems by linear ones.

There are two kinds of nonlinearities: continuous and discontinuous. A non linear system has a *continuous nonlinearity* if it can be described by a set of continuous functions and the control set is unconstrained. Linear approximations of such systems work well if the nonlinearity is not severe, and the state remains in a neighborhood of the initial point. However, the performance of the control law will degrade rapidly as the state evolves away from the initial point. In such systems, non linearities cause of poor control performance, and might even cause system instability.

A system has a *discontinuous nonlinearity* (often called *hard nonlinearity*) if it can be described by a set of discontinuous functions or the control set is constrained. Such systems cannot be linearized. They include all systems with friction, drag, saturation, or hysteresis, which are increasingly common in finance and economics.

Methods of nonlinear control

When applied to nonlinear systems, the maximum principle often yields an untractable system of equations. Without a practical general method of solution, a collection of methods have been developed to fit the various characteristics of the system.

A common design method for experienced control designers is simply "trial and error". The method requires thorough familiarity with the system, and the stability and robustness of the control is tested through simulations before full implementation. The trial and error method has two weaknesses: The method is appropriate for simple models, but gives poor results for larger, more complicated ones. The second weakness is that it is not possible to experiment with large systems in economics and finance, especially in the presence of uncertainty.

A second approach, called *feedback linearization*, is to choose a control with two components; the first one cancels the non linear elements of the model, and the second one to minimize the objective function, using linear control techniques. The method is very sensitive to modeling errors, and lacks robustness to parametric uncertainty, which explains why its successful use is limited to cases where the model can closely describe the system dynamics, mostly in physics and engineering.

A third approach is simply to approximate the model with a linear one in a neighborhood of some point in the state space. By doing so, the controller can use the many linear control techniques available, but cannot guarantee the stability or robustness of the system in a region outside the design neighborhood. Generally, linear approximations have been very successfully, which explains why the vast amount of research done in linear systems theory is not matched by an equal amount devoted to nonlinear systems. *Gain scheduling* in a natural extension of the state linearization technique: The controller designs a linear control law in the neighborhood of several points in the state space, and interpolates the corresponding feedback control laws between the points (almost all practical interpolations are either linear or splines). The composite control law performs quite well if enough points are chosen to cover the entire range of the state space. The major drawbacks for this technique are the heavy computational burden, and the need to model the system at several points in the state space..

A fourth approach is to use *robust* control, a combination of a control law optimizing an objective function, and a tracking control law aimed at keeping the state trajectory on the desired path, despite possible model uncertainties. The method implies a trade off between modeling accuracy and performance; Robust control is a sub-optimal control law (Which does not fully optimize the objective function) which yields a stable solution in the presence of parameter errors whereas conventional closed loop optimal control solutions can become unstable. Robust control is especially useful if the model parameters change rapidly and are not constantly re-estimated. The main drawback is that the control activity is often excessive (the tracking part of the control law requires frequent control action, or calls for a control action exceeding the admissible control bounds).

A fifth approach, *adaptive control,* incorporates control design with the estimation of parameters of an uncertain system. An enormous amount of research is being done in this area, most of it related to robotics and systems engineering[27]. Unlike robust control, adaptive control does not require that the controller obtain an estimate of the bounds of the control parameters. An important class of adaptive controllers is the "self tuning" controller, which rely on the *certainty equivalence* principle: The control process involves two steps; choose the control law using the system described with the latest parameter estimates available, and choose the adaptation law using the tracking error dynamics (the difference between expected and actual system performance) to update and forecast the system parameters. The method (which can be combined with robust control) is particularly suited for economic systems, where the model parameters can drift over time.

3.6 Observability and observers

The definition of observability for discrete systems is very similar to the definition for continuous systems introduced in chapter 2.

Observers

To determine what is the optimal control vector, we need to know the values of the state vector; Managers cannot determine the optimal production levels without knowledge of sales, and the Federal Reserve Bank cannot devise monetary policy without knowledge of inflation and interest rates. In reality, some of the needed information might not be available at the time decisions must be made, and proxies for the unavailable information must be substituted in the control law. Such proxies is called *observers.* The cost of not having complete and perfect knowledge of the relevant information at the time the control vector must be chosen can be substantial and controllers spend a large amount of time and resources estimating the missing information. Since all estimations include a stochastic element, the proxies are random variables, and we must delay the introduction of observers until after the introduction of stochastic control.

3.7 Economic examples

3.7.1 A partial adjustment multiplier-accelerator model of a closed economy

Kendrick's (1981) partial adjustment model of the US economy provides a simple illustration of an application of the tracking problem. Let C_k, I_k, and Y_k denote real consumption, investment and income in period k. Let C_k* and I_k* denote the desired levels of consumption and investment:

$$C_k^* = a + b\,Y_k$$

---

[27] A very comprehensive list of references on this topic can be found in: Slotine, J.J. and W. Li, (1991).

$$I_k^* = c + d\,(C_k - C_{k-1})\,,$$

and the adjustments of consumption and investment are described by:

$$C_k = C_{k-1} + e\,(C_k^* - C_{k-1})$$
$$I_k = I_{k-1} + f\,(I_k^* - I_{k-1})\,,$$

where e and f are partial adjustment parameters. Using the identity $Y_k = C_k + I_k + G_k$, and after substitution of the first two equations in the next two, we have the model:

$$C_k = \alpha_0 + \alpha_1 I_k + \alpha_2 C_{k-1} + \alpha_3 G_k$$
$$I_k = \beta_0 + \beta_1 C_k + \beta_2 C_{k-1} + \beta_3 I_{k-1}\,.$$

We can redefine the government expenditures G_k as the obligations approved in the previous period; $G_k = O_{k-1}$. The reduced form of the model was estimated with OLS, after substitution of O_{k-1} for G_k, and using quarterly data from 1947-Q2 to 1969-Q1 expressed in 1958 dollars;

$$\underset{}{C_k =} \underset{(.016)}{1.014\,C_{k-1}} + \underset{(.047)}{.002\,I_{k-1}} - \underset{(.031)}{.004\,O_{k-1}} \underset{(1.52)}{-1.312} \qquad R^2 = .998,\ \ DW = 2.19.$$

$$I_k = \underset{(.023)}{.093\,C_{k-1}} + \underset{(.068)}{.753\,I_{k-1}} - \underset{(.044)}{.100\,O_{k-1}} \underset{(2.16)}{-.448} \qquad R^2 = .938,\ \ DW = 1.62.$$

Using the method described in section 3.4b, we can absorb the constant terms and write the model as:

$$\begin{pmatrix} C \\ I \\ 1 \end{pmatrix}_{k+1} = \begin{bmatrix} 1.014 & .002 & -1.312 \\ .093 & .753 & .448 \\ 0 & 0 & 1 \end{bmatrix} \begin{pmatrix} C \\ I \\ 1 \end{pmatrix}_k + \begin{pmatrix} -.004 \\ -.100 \\ 0 \end{pmatrix} O_k\,,$$

which is now of the form (3.2.1). The desired path is a constant growth rate of .75% per quarter for consumption, investment and government obligations over the next 7 periods. Starting from the end of the estimation period, where C_0=460.1, I_0=113.1, O_0=153.6, the desired state and control vectors are: $x_k^* = ((1.0075)^k 460.1, (1.0075)^k 113.1, 1)'$ and $u_k^* = (1.0075)^k(153.6)$. In this model, the elements r_k and C_k of (3.2.38) are the desired state vector and the identity matrix respectively. The objective is to minimize deviations of the state and control vectors from the desired path. Hence we choose the quadratic objective function:

$$J=\left(x_T-x_T^*\right)' P_T\left(x_T-x_T^*\right)+\sum_{k=1}^{T}\left[\left(x_k-x_k^*\right)' Q_k\left(x_k-x_k^*\right)+\left(u_k-u_k^*\right)' R_k\left(u_k-u_k^*\right)\right]$$

with P_T = diag(100, 100, 0), Q_t = diag(1, 1, 0), and R_k=1, and T=7. Note that the penalty for not meeting the desired values of the state vector in the final period is 100 times greater than before that period. Since the objective function is expressed in deviation form, we must rewrite the state equation in terms of the deviations from the desired paths, which also introduces more constants. Again, we absorb the constants using section 3.4b. Finally, after a few manipulations, equations (3.2.43) to (3.2.50) yield the state and optimal control trajectories. The results are summarized in the following table, where the lower case letters represent the absolute value of the difference between the actual and desired state variables.

k :	1	2	3	4	5	6	7
c_k :	1.2	2.6	4.0	5.3	6.8	8.3	9.8
i_k :	1.1	1.9	2.2	2.3	2.1	1.6	.9
C_k:	464.8	469.6	474.5	479.4	484.4	489.5	494.6
I_k:	112.8	112.9	113.4	114.2	115.3	116.7	118.3
O_k:	156.4	156.8	157.2	157.4	157.2	156.7	155.6

Note that the poor performance at keeping consumption on the desired path is not surprising since the coefficient of C_{k-1} is greater than unity, which implies that the model has an explosive behavior. The control sequence does keep the investment close to the desired level, with a closer fit during the last period when the penalty is the highest. This partial success in controlling investment does not help much in controlling consumption because the coefficient of investment in the first equation is so small. The mediocre results point to two possible problems with the model; the system might not be controllable (with only one instrument and two targets), or the model does not capture some important relationship between the variables. For some interesting extension and analyses of this model (in a stochastic environment), the reader should consult Kendrick (1979) and (1981).

3.7.2 The expectations-augmented Phillips relation

In this example we show how a continuous time model can be represented by an equivalent discrete time formulation. We use a very simple model of the Phillips relation, where the nominal wage demands are a function of the expected rate of inflation. The assumed relationship between the actual inflation rate and the unemployment rate is:

$$p = \alpha_1 - T - \alpha_2 U + \alpha_3 \pi$$

where p is the actual rate of inflation, T is the exogenous percentage increase in labor productivity, U is the unemployment rate, π is the expected rate of inflation, α_1, α_2 and α_3 are parameters. The dynamics of the model is provided by;

$$d\pi/dt = j\,(p - \pi) \quad (0 < j < 1).$$

Monetary policy is the only policy instrument of this continuous time model. The desired target is to reach a steady state where the actual unemployment rate is equal to the natural level of unemployment, and where both the expected and the real rate of inflation are equal to zero. The rate of change of unemployment is assumed proportional to the difference between the growth rate of nominal money supply m, and of the actual rate of inflation;

$$dU/dt = -k(m - p), \qquad \text{with} \quad k > 0.$$

The three equations above constitute a closed model in the three variables; π, p and U. A few manipulations[28] lead to a differential equation describing the time path of expected inflation:

$$d^2\pi/dt^2 = -\,j\alpha_2 k\pi - [\alpha_2 k + j(1-\alpha_3)]d\pi/dt + j\alpha_2 km$$

After the change of variables $x_1=\pi$, $x_2=d\pi/dt$, $u=m$, $a_1=-j\alpha_2 k$, and $a_2=-\alpha_2 k+j(1-\alpha_3)$, the equation above can be written in the usual notation: $dx/dt = Ax + Bu$, where

$$x = \begin{pmatrix} \pi \\ d\pi/dt \end{pmatrix}, \qquad A = \begin{bmatrix} 0 & 1 \\ a_1 & a_2 \end{bmatrix}, \qquad B = \begin{pmatrix} 0 \\ -a_1 \end{pmatrix}$$

We assume that the objective of the policy maker is to minimize the function

$$J = \int_{t_0}^{\infty} [x'Qx + u'Ru]dt$$

where Q and R are both positive definite diagonal matrices (the elements of the diagonal of Q represent the relative weights attached to the dual objective of reducing expected inflation and reducing the change in the expected rate of inflation). Since the ultimate objective is to reduce the expected rate of inflation to zero, a greater weight is assigned to the first diagonal element of Q. The element R determines how the control action influences the objective function, i.e. the cost of

---

28 Substitute the first equation in the second, differentiate the result with respect to time, and eliminate p using the third equation.

changing the growth rate of the real money supply. For this example, we arbitrarily set $Q = \text{diag}(1, 0.2)$, and $R = 0.5$. As formulated. the model is an infinite horizon continuous time linear quadratic regulator problem.

A discrete version of the state equation was estimated with restricted OLS, using non seasonally adjusted monthly data for the US economy. Changes in the producers price index and M1 were used as a proxy for expected inflation and for the money supply, respectively. The estimated equation is:

$$d^2\pi/dt^2 = \underset{(.047)}{-.284}\,\pi + \underset{(.051)}{1.644}\, d\pi/dt + \underset{(.047)}{.284}\, m \qquad R^2 = .7746$$

where the numbers in parenthesis are standard errors. All the regression parameters are significant at the .01% level. With $a_1 = -.284$ and $a_2 = 1.644$, the matrix A has eigenvalues $\mu_1 = -1.178$, and $\mu_2 = -.465$. To enhance the probability that the system is stable, and to illustrate the use of the coefficient of relative stability (described in section 3.4c), we choose $\theta = .5$, where θ is the real and non negative coefficient of relative stability.

The solution of this linear quadratic regulator problem defined is given by $u_\theta(t) = -K_\theta(t)x(t)$, where $K_\theta = R^{-1}B'P_\theta x(t)$, and P_θ is the unique positive semidefinite solution of the algebraic Riccati equation:

$$(A+\theta I_n)'P_\theta + P_\theta'(A+\theta I_n) - P_\theta BR^{-1}B'P_\theta + Q = 0 .$$

Note that since we use a discrete representation of a continuous model, we also use the Riccati equation of the continuous formulation (chapter 2). The closed loop state feedback matrix $A_{cl\theta} = (A+\theta I_n - BK_\theta)$, completely determines the optimal path of the state variables from given initial conditions. With the parameter values estimated above, we have:

$$P_\theta = \begin{bmatrix} 54.43 & 16.86 \\ 16.86 & 32.96 \end{bmatrix}, \quad K_\theta = \begin{pmatrix} 9.478 \\ 18.68 \end{pmatrix}, \quad \text{and} \quad A_{cl\theta} = \begin{bmatrix} 0 & 1 \\ -2.975 & -3.662 \end{bmatrix}.$$

The eigenvalues of the matrix $A_{cl\theta}$ are $\mu_1 = -2.445$ and $\mu_2 = -1.216$, and since both are negative, we can conclude (from chapter 2) that the system is asymptotically stable under the control law.

3.7.3 Optimal control of a closed economy with bond-financed budget deficit

Let Y_t and G_t denote national income and government expenditures in period t, let B_{t-1} denote interest payment in period t for government bonds issued in the preceding periods, and let Y_t^*, G_t^*, and B_{t-1}^* denote the corresponding levels desired by a policy maker. R_t is the real interest rate, and M_t is the stock of high power money. Murata (1982) proposes the following simplified model of a closed economy:

$$Y_t = a_1 (Y_t + B_{t-1}) - a_2 R_t + G_t$$
$$M_t = a_3 Y_t - a_4 R_t$$
$$\Delta M_t - R_t^{-1} \Delta B_t = G_t - a_5 Y_t + a_6 B_{t-1}$$

The first equation is simply the sum of consumption $(a_1 (Y_t + B_{t-1}))$, investment expressed as a linear function of the interest rate $(- a_2 R_t)$, and government expenditures. The second equation is a linearized liquidity preference function, and the third equation is the constraint that the government finances its deficit by increasing the money supply and issuing new bonds. Following Murata we express the reduced form of the model in terms of deviations from the desired levels:

$$y_{t+1} = \alpha_1 y_t + \alpha_2 g_t + \alpha_3 b_t + \alpha_4 b_{t-1} + \beta_1 \Delta g_t + \beta_2 \Delta b_t$$

where $y_t = (Y - Y^*)$, $g_t = (Y - Y^*)$, $b_t = (Y - Y^*)$ are the state variables, and the planned changes in deviations $\Delta g_t = (g_{t+1} - g_t)$, $\Delta b_{t-1} = (b_{t+1} - b_t)$ are the control variables. Estimates of the parameters are: $\alpha_1 = .6875$, $\alpha_2 = 1.5625$, $\alpha_3 = 2.3125$, $\alpha_4 = -1.0625$, $\beta_1 = 3.125$, $\beta_2 = -31.25$.
Using the definition of the control variables and the identity $b_{t-1} = b_{t-1}$, we can write the model as:

$$\begin{pmatrix} y_{t+1} \\ g_{t+1} \\ b_{t+1} \\ b_t \end{pmatrix} = \begin{bmatrix} .6875 & 1.5625 & 2.3125 & -1.0625 \\ 0 & 1 & 0 & 0 \\ 0 & 0 & 1 & 0 \\ 0 & 0 & 1 & 0 \end{bmatrix} \begin{pmatrix} y_t \\ g_t \\ b_t \\ b_{t-1} \end{pmatrix} + \begin{bmatrix} 3.125 & -31.25 \\ 1 & 0 \\ 0 & 1 \\ 0 & 0 \end{bmatrix} \begin{pmatrix} \Delta g_t \\ \Delta b_t \end{pmatrix}$$

which is of the form (3.2.1), with the state vector $x_t = (y_t , g_t , b_t , b_{t-1})'$ and the control vector $u_t = (\Delta g_t , \Delta b_t)'$. The objective function is assumed quadratic of the form (3.2.2), where $S_T = \text{diag}(1, 0, \gamma, 0)$, $Q_t = \text{diag}(1, 0, \varepsilon, 0)$, and $R_t = \text{diag}(\phi_1, \phi_2)$. With $\gamma > 0$ $\varepsilon > 0$ $\phi_1 > 0$ and $\phi_2 > 0$, the matrix R_t is positive definite, and S_T, Q_t are positive semidefinite. If we set the planning horizon $T = 5$, the control problem is now the finite horizon free final state problem defined in case 1 of section 3.2. Using the results of that section, we can investigate the behavior of the model for different values of the parameters γ, ε, ϕ_1, ϕ_2 of the objective function. We arbitrarily set the initial condition $x_0 = (-10, 0, 0, 0)'$ (i.e. we assume that at the start of the control periods, government expenditures and interest payment on bonds are exactly equal to their desired values, while the income level is 10 units less than the desired level).

The behavior of an optimal control model depends critically upon the parameters of the objective function. For some economic problems, like the maximization of a profit function, these parameters are defined by the model. However, most macroeconomic models have no easily identifiable objective function and the policy maker is often reduced to assign a value to these parameters rather arbitrarily, using their own experience or perception of the problem at task. It is therefore very important to investigate the behavior of the model for a range of values of the chosen parameters to ensure that the model performs in the desired manner.

Model performance with $\gamma = \varepsilon = 2$, $\phi_1 = .5$ $\phi_2 = 1$;
With the chosen parameters, there is an equal penalty for not meeting the income and bond payment target, but no penalty for failing to meet the government expenditures target. The penalty for not meeting the targets is independent of time[29]. The parameters ϕ_1 and ϕ_2 represent the cost of using the controls. The optimal control sequence is obtained using the relation (3.2.31), where the Kalman gain matrices K_t and the solutions of the Riccati equations S_t are obtained recursively, backward in time, starting from the final state. We first compute the sequence of Kalman gain matrices (3.2.32):

$$K_1 = \begin{bmatrix} .041 & 276 & 1.837 & -.064 \\ -.018 & .022 & .113 & .027 \end{bmatrix} \qquad K_2 = \begin{bmatrix} .039 & .230 & 1.739 & -.061 \\ -.018 & -.027 & .103 & .028 \end{bmatrix}$$

$$K_3 = \begin{bmatrix} .034 & .160 & 1.453 & -.052 \\ -.019 & -.034 & .074 & .029 \end{bmatrix} \qquad K_4 = \begin{bmatrix} -.024 & .084 & 964 & -.037 \\ -.020 & -.041 & .025 & .030 \end{bmatrix}$$

$$K_5 = \begin{bmatrix} .012 & .028 & .418 & -.019 \\ -.021 & -.047 & -.030 & .032 \end{bmatrix}$$

The corresponding control sequence, with the associated value of y_t is:

t:	1	2	3	4	5
Δg_t:	.412	.216	.107	.041	.007
Δb_t:	-.178	.029	.037	.038	.038
y_t :	-.027	-.015	-.009	-.005	-.001

Model performance with $\gamma = \varepsilon = 0$, $\phi_1 = .5$ $\phi_2 = 1$;

29 In macroeconomic models, the objective function chosen usually imposes a much larger penalty for failing to meet the target in the final period. That choice is based on the political cost of not meeting the public's expectations in an election year. The example of section 3.7.1 provides an example of such an objective function.

In this simulation, the cost of using the controls remains unchanged, but there is no penalty for not meeting the target for bond payment. The corresponding control sequence, with the associated value of y_t is:

t:	1	2	3	4	5
Δg_t:	.047	.005	.003	.002	.001
Δb_t:	-.215	-.013	-.0067	-.0067	-.0069
y_t :	-.0066	-.0003	-.0001	-.0001	.0002

Comparison of the control actions between the two cases confirm the intuitive reasoning that if the system is subject to less constraints, it must be easier to control. Indeed, by comparing the two tables above, it is clear that it takes less control action in the second case to keep the income on the desired path, and that convergence to the desired path is much faster than in the first case.

References

Arrow, K. and Kurz, M, (1971), *Public investment, the rate of return, and Optimal Fiscal Policy*. Baltimore: John Hopkins Press.

Friedman, M., (1968). "The role of Monetary Policy", *American Economic Review*, 1-17

Kendrick, D., (1981), *Stochastic control for economic models*, McGraw Hill.

Kendrick, D., (1979), "Adaptive Control of Macroeconomic Models with Measurement Error", in *Optimal Control for Economic Models*, Holly, Rüstem and Zarrop editors, St. Martin's Press, pages 204-227.

Murata, Y., (1982), *Optimal Control Methods for Linear Discrete-Time Economic Systems*, Springer-Verlag.

Patel, R. and Mundo, N. (1982) "Multivariable System Theory & Design", Pergamon Press.

Slotine, J.J. and W. Li, *Applied Nonlinear Control*, Prentice Hall, 1991.

Wang, Y., (1989), "A Note on the Solution of the Algebraic Riccati Equation", *Systems & Control Letters*, 12(5), pages 465-472.

CHAPTER 4: **Stochastic Control Theory**

4.0 Introduction

Methods of applied stochastic control may be viewed as extensions of the deterministic control approach in two ways. One is the characterization of the stochastic process over time that may generate the deterministic control model. The second is the estimation aspect where the parameters of the deterministic control model are assumed to be unknown, and time series data are used for estimating those parameters.

We would discuss these two aspects in a simplified framework by means of concrete examples and their economic implications. We start with a brief introduction to the theory of stochastic processes. This is followed by economic models which derive optimal decision (or control) rules under linear quadratic game theoretic and dynamic portfolio models.

4.1 Stochastic processes under control

A stochastic process may be viewed in two ways. One is that it is simply a probability process; that is, any process in nature whose evolution can be analyzed successfully in terms of probability, i.e., the transition probability. A second view is that it is a family of random variables denoted by x_t or $x(t)$ indexed by $t \in T$. The variation of t over the index set T introduces a dynamic element in the behavior of the random variable, which can be analyzed in two interrelated ways: either in terms of the probability functions dependent on t as t varies, or in terms of the expected value or variance function of $x(t)$ as t changes. One useful way of division of stochastic processes is in terms of four classes, taking the case where we have one random variable at any particular t: (i) both x and t discrete, (ii) x discrete and t continuous, (iii) x continuous and t discrete and (iv) both x and t continuous. In each of the cases above one can define appropriate probability density functions. Stochastic processes can be classified in several ways depending on the criteria chosen. Three most common criteria are (a) the values taken by the random variable $x(t)$ being in the domain of positive integers or not, (b) the extent of dependence or independence of $x(t)$ on past history or memory and (c) the invariance or otherwise of the joint distribution function of the collection of random variables $x_1 = x(t_1)$, $x_2 = x(t_2), \ldots, x_n = x(t_n)$ under an arbitrary translation of the time parameter.

If the process $x(t)$ is capable of assuming positive integral values only, then it is known as point processes, for in such a case, $x(t)$ can be identified with events or incidences represented as points on the time axis t.

The second criterion introduces the classification of Markovian and non-Markovian processes. For any parametric value t and arbitrarily small $h > 0$, the probability structure of the process corresponding to the parametric value $(t+h)$ may depend only on the probability structure at t. In such a case $x(t)$ is called a Markov process; non-Markovian processes then consist of those processes which do not satisfy the above property of the Markov process known as the Markov property. An alternative way to define Markovian processes is through the joint distribution function of the random variables $X_1 = X(t_1)$, $X_2 = X(t_2)$, ... , $X_n = X(t_n)$ for any finite set of $(t_1, t_2, \dots, t_n)$ of $t \in T$, T being the index set which is denoted as:

$$F_n(x_1, t_1; x_2, t_2; \dots; x_n, t_n) = F_{X_1, X_2, \dots X_n}(x_1, t_1; x_2, t_2; \dots; x_n, t_n)$$

$$= P[X_1 \le x_1, X_2 \le x_2, \dots X_n \le x_n] \tag{4.1.1}$$

where $x_i = x(t_i)$ and P denotes the joint probability. Thus the process $\{X(t), t \in T\}$ is Markovian, if for every n and for $t_1 < t_2 \dots < t_n$ in the index set T we have

$$F_n(x_n, t_n | x_{n-1}, t_{n-1}; x_{n-2}, t_{n-2} \dots; x_1, t_1) = F_n(x_n, t_n | x_{n-1}, t_{n-1}) \tag{4.1.2}$$

If the probability density function f of F exists, then (4.1.2) is equivalent to

$$f_n(x_n, t_n | x_{n-1}, t_{n-1}; x_{n-2}, t_{n-2} \dots; x_1, t_1) = f_n(x_n, t_n | x_{n-1}, t_{n-1})$$

On the basis of this relation an important equation known as the Chapman-Kolmogorov equation can be derived.

The invariance criterion classifies stochastic processes into stationary and nonstationary. A process $\{X(t), t \in T\}$ is defined to be strictly stationary if for each n and for every arbitrary τ we have for the distribution function $F_n(\cdot)$ defined in (4.1.1)

$$F_n(x_1, t_1; x_2, t_2; \dots; x_n, t_n) = F_n(x_1, t_1+\tau; x_2, t_2+\tau; \dots; x_n, t_n+\tau) \tag{4.1.3}$$

For $t_j + \tau \in T$, $j = 1, 2, \dots, n$ identically. Thus the family of joint probability distributions remains invariant under an arbitrary translation of time. Thus it follows that the distribution function $F_1(x,t)$ is not a function of t and

$$F_2(x_1, t_1; x_2, t_2) = F_2(x_1, x_2; t_2 - t_1)$$

Hence if the moments exist we would have:

$E[X(t)]$ = constant, independent of time

$E[X(t)X'(t+\tau)] = R(\tau) = R(-\tau)$

i.e., covariance of $X(t)$ and $X(t+\tau)$ do not depend on time t but depend only on the lag τ.

A weaker concept than strict stationarity is provided by second-order stationary. Thus a process $\{X(t),\ t \in T\}$is wide-sense, or weakly or *covariance*, or *second-order stationary* if
$|E[X(t)]|$ is a finite constant,
$E[X(t)^2]$ is bounded,
$E[X(t_1)X'(t_2)] = R(t_2\text{-}t_1)$.
Stationarity can also be defined by a process having a constant spectral density.

In economic applications Markov processes which are stationary in some sense have played very important parts. Reasons are several. First, the practicing econometricians had found very early that autoregressive systems with distributed lags fit very well several kinds of economic behavior, e.g., cobweb cycle, distributed lags, capital-stock adjustment. Second, use of the Chapman-Kolmogorov equation showed that differential equations representing dynamic behavior may be interpreted as dynamic random equations, of which the solutions or trajectories may be interpreted probabilistically. Since the solutions of random differential equations may or may not converge to their steady state equilibrium values, depending on certain conditions, once could obtain a generalized view of stochastic equilibrium or disequilibrium. Thus for instance, a deterministic differential equation may have a convergent solution as time approaches infinity, while a stochastic analogue may not. Third, the Markovian property is very useful in modeling expectation formations and learning and even for non-Markovian situations the Markovian assumption is a good first approximation, especially when empirical data is very short.

One of the most useful applications of Markov processes is in terms of increment processes, defined over continuous time. Consider the stochastic process $\{X(t)\}$ for $t \geq 0$. Denote the difference $X(t_2)\text{-}X(t_1)$ by $X(t_1,t_2)$ where $t_2 > t_1 > 0$, which is termed an increment of $X(t)$ on $[t_1,t_2]$. If for all $t_1 < t_2 < \ldots < t_n$, the successive increments $X(t_1,t_2), X(t_2,t_3), \ldots, X(t_{n-1},t_n)$ are mutually statistically independent, then the process $\{X(t), t \geq 0\}$ is called an *independent-increment* stochastic process. If for this process the probability distributions of its increments $X(t_1,t_2), X(t_2,t_3), \ldots, X(t_{n-1},t_n)$ depends only on the parameter differences $t_2\text{-}t_1$, $t_3\text{-}t_2, \ldots, t_n\text{-}t_{n-1}$ then the process $X(t)$ is said to have *stationary independent increments.* Three important examples of such processes which have found widest applications are (a) the *Brownian motion*, named after its discoverer Robert Brown, a botanist; this is also called Wiener process or Wiener-Levy process that was widely applied by Einstein, (b) the *Poisson process*, which is most widely applied in queuing, telephone traffic and other fields of operations research and (c) *Gaussian stochastic process*, which is most frequently applied due to its analytic properties and the fact that the central limit theorem gives it a wider applicability.

More formally a Brownian process $\{X(t), t \geq 0\}$, which may now be denoted by $X(t,w)$, $t \in T$, $w \in W$ to indicate its dependence on the sample space W is a process satisfying four conditions: (i) $X(0,w) = 0$ by convention, i.e., the process starts at zero, (ii) stationary independent increments, (iii) the increment $X(t) - X(s)$, $t > s > 0$ has normal or Gaussian distribution with zero mean and variance $\sigma^2(t\text{-}s)$, and (iv) for each sample index $w \in W$ the process $X(t,w)$ is continuous in t for $t \geq 0$. The most significant contribution of Wiener on the development of the theory of Brownian motion is to show that the sample functions $X(t,w)$ viewed as functions of $w \in W$ are continuous but not differentiable.

The second example of an independent increment stochastic process is the Poisson process. The only difference here is that the process $X(t)$, $t \geq 0$ has independent integer-valued increments. Arrival of customers in queue, sequence of production and inventories, spread of impulse noises are modeled by the Poisson process. Estimation of parameters of Poisson processes applied to economic models and its implications are discussed by Tintner and Sengupta (1972).

Lastly, a stochastic process $\{X(t), t \in T\}$ is called Gaussian, if for every finite set $t_1, t_2, .., t_n$ the random variables $X(t_1), X(t_2), \ldots ,X(t_n)$ have a joint normal distribution with a mean vector $m(t) = (E[X(t_1)], E[X(t_2)], \ldots, E[X(t_n)])'$ and variance-covariance matrix $V = (v_{ij})$, where $v_{ij} = E[(X(t_i)\text{-}m(t_i))(X(t_j)\text{-}m(t_j))']$. In the stationary case the mean vector $m(t)$ is a constant (i.e., time-independent) and the covariance matrix is a function of the time instants only through their differences.

A concept associated with Gaussian processes that is most useful in interpreting random differential equation is the white noise property. Loosely speaking, a white noise process is defined by the derivative of a Brownian or Wiener process. But a Wiener process is not differentiable in the mean square sense, since by its definition given before we have

$$E\left[\frac{X(t)-X(s)}{t-s}\right]^2 = \left(\frac{s^2}{t-s}\right),$$

which becomes infinite as $(t\text{-}s)$ approaches zero. Hence one has to proceed differently. Let $W(t)$, $t \geq 0$ denote the Wiener process, which is by definition Gaussian with zero mean and covariance denoted by $\mu(t,s) = 2(t\text{-}s)\sigma$, where $\sigma^2 = c$ is a finite constant and $t > s$. The formal derivative of $W(t)$, denoted by $\dot{W}(t)$ is clearly Gaussian with mean zero and covariance given by

$$\mu_{\dot{W}(t)}(t,s) = \frac{\partial^2 \mu(t,s)}{\partial t\, \partial s} = 2c\frac{\partial^2 \min(t,s)}{\partial t\, \partial s} = 2c\delta(t-s), \tag{4.1.4}$$

where $\delta(t-s) = \begin{cases} 0, & if \quad t \leq s \\ 1, & if \quad t > s \end{cases}.$

The most important use of Gaussian white noise processes is in the theory of random differential equations, where Ito's theorem has to be used for taking account of the nondifferentiability property of the white noise process. Taking the linear scalar case we have the dynamic system

$$\dot{x}(t) = a(t)x(t) + b(t)w(t), \qquad x(t_0) = x_0 \tag{4.1.5}$$

where $a(t)$, $b(t)$ are non random time-dependent coefficients and $w(t)$ is Gaussian white noise independent of x_0. One could write (4.1.5) alternatively as

$$dx(t) = a(t)\, x(t)\, dt + b(t)\, dn, \qquad x(t_0) = x_0 \tag{4.1.6}$$

where we have used $dn(t)/dt = W(t)$ or, on integration

$$x(t) - x(t_0) = \int_{t_0}^{t} a(s)x(s)ds + \int_{t_0}^{t} b(s)dn(s), \tag{4.1.7}$$

with $x(t_0) = x_0$. Whereas the first integral on the right-hand side of (4.1.7) is well defined as a Riemann integral, the second one may not since the random variable y_n defined by the partial sum

$$y_n = \sum_{i=1}^{k} x(t_i)\left(n(t_i) - n(t_{i-1})\right) \tag{4.1.8}$$

does not converge in the mean square sense to a unique limit. Ito's theorem provided a method of selecting the subdivisions $\{t_i, t_i\text{-}t_{i-1}\}$ so that the limit of y_n in (4.1.8) exists as $\Delta_m = \max(t_{i+1}\text{-}t_i)$ approaches zero.

For specific applications of Ito's rules of stochastic calculus to any functional $z(t,x(t))$ defined on $x(t)$ satisfying (4.1.6) we proceed as follows: we define the differential of $Z(t) = z(t,x(t))$ as

$$dZ(t) = \left[z_t + z_x a(t) + \tfrac{1}{2} z_{xx} \left(b(t)\right)^2 \right] dt + z_x b(t) dn(t) \tag{4.1.9}$$

where $z_t = \partial z(t,x(t))/\partial t$, $z_x = \partial z(t,x(t))/\partial x$, and $z_{xx} = \partial^2 z(t,x(t))/\partial x^2$. This differential exists under certain regularity conditions. In general if $Z(t) = z(t,x_1(t), x_2(t), \ldots, x_m(t))$ is a continuous function with continuous partial derivatives z_t, z_{x_i} and $z_{x_i x_j}$ and we have in place of (4.1.6) the stochastic differentials on $[t_0,T]$

$$dx_i(t) = a_i(t)\, dt + b_i(t)\, dn(t), \qquad i = 1, 2, \ldots, m$$

then $Z(t)$ also possesses a stochastic differential in the same interval given by:

$$dZ(t) = z_t + \sum_{i=1}^{m} z_{x_i} dx_i + \tfrac{1}{2} \sum_{i,j=1}^{m} z_{x_i x_j} dx_i dx_j \tag{4.1.10}$$

where $dx_i dx_j = b_i(t)b_j(t)dt$, $i,j \le m$. Some examples of applying the Ito differential rules are as follows: Let $Z(t) = x_1(t)x_2(t)$ and

$$dx_1(t) = a_1(t)dt + b_1(t)dn(t)$$

$$dx_2(t) = a_2(t)dt + b_2(t)dn(t)$$

where $dn(t) = W(t)\,dt$ is the Gaussian white noise. On applying (4.1.10) one gets

$$d(x_1(t)x_2(t)) = x_1(t)dx_2(t) + x_2(t)dx_1(t) + b_1(t)b_2(t)dt$$

$$= [x_1(t)a_2(t) + x_2(t)a_1(t) + b_1(t)b_2(t)]dt + [x_1(t)b_2(t) + x_2(t)b_1(t) + b_1(t)]dn(t).$$

Take another scalar case where $Z(t) = z(x_0, e^{at})$ where a is a constant and $dx(t) = ax(t)dt$. Then $dZ(t) = z_t dt + z_x dx + \frac{1}{2} z_{xx} dx dx = ax_0 e^{at} dt$. By applying these rules the solution of the linear scalar random differential equation (4.1.6) taken in a more general form as

$$dx(t) = a(t)x(t) + \sum_{i=1}^{m} b_i(t)x(t)dn_i(t), \tag{4.1.11}$$

where $x(t_0) = c$, $dn(t) = \{dn_i(t)\}$ = m-dimensional white noise, can be explicitly written as

$$x(t) = c\exp\left[\int_{t_0}^{t}\left(a(s) - \sum_{i=1}^{m}\tfrac{1}{2}b_i^2(s)\right)ds + \sum_{i=1}^{m}\int_{t_0}^{t} b_i(s)dn_i(s)\right]$$

of $a(t)$, $b_i(t)$ are constants, then we have

$$x(t) = c\exp\left[\left(a - \sum_{i=1}^{m}\tfrac{1}{2}b_i^2\right)(t - t_0) + \sum_{i=1}^{m} b_i\left(n_i(t) - n_i(t_0)\right)\right] \tag{4.1.12}$$

This result (4.1.12) is very useful for economic applications.

4.2. Economic applications

4.2.1 Example 1: Feedback Rules

Recent economic applications of control theory have spanned a wide spectrum. Starting from linear feedback rules in stabilization policies in a Keynesian dynamic model, it has moved into areas of differential games and adaptive control where the risk attitudes if the various agents play a very critical role. Models of rational expectations which involve the dynamic interaction between future expectations and their realizations have also played a central role in the estimation of dynamic adjustment models, which frequently use the linear decision rule approach.

We now consider several economic examples to illustrate the various modern applications of control theory. To start with a simplest case, consider a proportional feedback rule of government policy analyzed by Phillips (1954) under a Keynesian model

$$dy(t) = ay(t) + dg(t) = ay(t) + by(t)dn(t)$$

where $dy(t)$ is deviations of real income from the target level, the government policy intervention $g(t)$ is of the form $dg(t) = by(t)dn(t)$, with $dn(t)$ being a white noise Gaussian process and a,b are real constants. The income solution $y(t)$ is now given by

$$y(t) = c \exp[(a - \tfrac{1}{2} b^2)(t - t_0) + b(n(t) - n(t_0)) .$$

Assuming that the law of large numbers holds for the $n(t)$ process, we would have $\lim y(t)$ equal to zero as t becomes infinite only if $a < \tfrac{1}{2} b^2$; otherwise the controlled system would be unstable. In his original work Phillips argued in terms of only a deterministic differential equation systems and showed that in many cases the government policy itself may be destabilizing. Here we find additional stochastic reasons why this may be so. Next we consider another version of Phillips' model involving stabilization policy. This is a multiplier-accelerator model with the government expenditure as a control variable. This may be written as

$$(aD^2 + bD + c)y_t - g(t) = \varepsilon_t \tag{4.2.1}$$

where D is the differential operator, g_t is government expenditure and the other constant parameters a, b and c may be related to the multiplier-accelerator mechanisms of Keynesian models and ε_t is a white noise stochastic process. Now consider this second-order model in a slightly different form, when the proportional and derivative policy of government are incorporated:

$$[D^2 + b_2D + (b_1 + e_{1t})]y_t = \varepsilon_{2t} \tag{4.2.2}$$

where the fixed coefficients are: $b_1 = (c + k_1)/a$, $b_2 = (b + k_2)/a$ and e_{1t}, e_{2t} (where e_{2t} contains ε_t and other shocks due to g_t) are assumed to be a two-dimensional Wiener process with zero mean and finite covariances $E[e_{it}e_{jt}] = 2B_{ij}dt$. It is known that this type of differential equation model satisfies its moment equations in variables $m_{ij}(t) = E[y^i_{1t}y^j_{2t}]$ where $y_{1t} = y_t$, $y_{2t} = Dy_t$ and the first-order moments satisfy

$$Dm_{10}(t) = m_{10}(t) \tag{4.2.3}$$

$$Dm_{01}(t) = - b_1 m_{10}(t) - b_2 m_{01}(t)$$

the second-order moments also satisfy the differential equations

$$Dm_{20}(t) = 2m_{11}(t)$$

$$Dm_{11}(t) = - b_1 m_{20}(t) - b_2 m_{11}(t) + m_{02}(t) \tag{4.2.4}$$

$$Dm_{02}(t) = 2\{b_{11} m_{20}(t) - b_1 m_{11}(t) - b_2 m_{02}(t) + B_{22} - 2B_{12} m_{10}(t)\}$$

Whereas the solutions for equations (4.2.3) specify the conditions for stability in the mean, those for (4.2.4) indicate the conditions for mean square stability. It is clear from the characteristic equation of (4.2.3) that its roots have negative real parts if the coefficients b_1, b_2 are positive, i.e., this result which specifies the requirements for asymptotic mean stability is the same as in the deterministic version. Assuming positive b_1, b_2, we may solve the second-order moment equation (4.2.4) for analyzing the asymptotic stability in mean square. The characteristic equation is

$$\phi(\lambda) = \lambda^3 + 3b_2\lambda^2 + 2(b_2^2 + 2b_1)\lambda + 4(b_1 b_2 - B_{11}) = 0.$$

Applying the Routh-Hurwitz stability criterion, it is clear that the conditions for asymptotic mean square stability are

(i) $b_1 > 0$, $b_1 > 0$, (which are conditions for mean stability),

(4.2.5)

(ii) $b_1 b_2 > B_{11}$.

Thus the second condition of (4.2.5) implies additional restrictions on the choice of the strength parameters k_1, k_2 of the stabilization policy. Thus under the prescribed control policy, mean income may converge to the target level but not necessarily with decreasing variance.

This apparent conflict between the mean and variance of the target variables {e.g. income in this case) when a certain type of control strategy is adopted appears in many dynamic economic models e.g., portfolio models of investment. Recently Holly and Hallett (1989) have discussed this aspect in terms of two types of optimal control or decision rules; one is the economist's approach, ant the other is the statistician's. The former approach combines the mean and the variance of the target variables in a transformed objective function and then derives the optimal control rules. These rules hold the high risk variables closer to their ideal paths than would otherwise be the case. Then the random shocks represented by ε_t would cause smaller variations about those paths. The statistician's approach is to choose optimal decision rules which are *robust* (i.e. less sensitive) in relation to *ex post* optimal path of the target variables. Thus the statistician's approach is designed to reduce the sensitivity of the optimal decisions themselves to the potential uncertainties from the outside. Thus the economist's approach is to choose *stable* targets but flexible instruments (or controls), while the statistician tends to choose flexible targets and sticky (or fixed) instruments. Both are possible responses for the uncertainty problem.

4.2.1 Linear Decision Rules

The linear decision rule (LDR) arises very frequently in applied cases known as the linear quadratic Gaussian model, where the objective function is quadratic in the state and control variables, and the model dynamics is linear with a disturbance term which is independently distributed The LQG model in its simple version can be written as:

$$\text{Minimize } J = E_t\left[\sum_{t=0}^{N-1}\left(x'(t)Qx(t) + u'(t)Qu(t)\right)\right]$$

$$\text{subject to: } \quad x(t+1) = Ax(t) + Bu(t) + \varepsilon(t) \tag{4.2.6}$$

where $x(t)$ and $u(t)$ are the state and control vectors, $x(0) = c$ is a fixed constant, where it is assumed for simplicity that the matrices Q, R, A, and B are constant parameters and the disturbance vector $\varepsilon(t)$ is distributed normally and independently of the controls $u(t)$ with zero means and fixed covariances. For this to define a strictly convex program, we would require the matrices Q and R to be positive definite. In most economic applications, $x(t)$ and $u(t)$ are measured as deviations $(x(t) - x^0(t))$ and $(u(t) - u^0(t))$ from known desired levels $(x^0(t), u^0(t))$, with Q and R being identity matrices. This type of LDR approach has several flexible features which may be briefly mentioned here.

First of all, this problem can be solved in two ways. Either by applying the standard necessary conditions for optimization of a quadratic programming model, or by dynamic programming. For the later method, we start with the last period N-1, substitute the linear dynamic equation into the objective function $J = J(\text{N-1})$, and minimize with respect to the control vector $u(t)$. The result yields the LDR which is the optimal feedback rule:

$$u^*(\text{N-1}) = Kx(\text{N-2}) + k$$

where: $K = -(R + B'QB)^{-1}B'QA$

$$k = -(R + B'QB)^{-1}Q\, E[\varepsilon(t)] = 0.$$

For any intermediate time period τ where N-1 can be period τ, the LDR solution would be of the same form i.e.,

$$u^*(\tau) = Kx(\tau\text{-}1) + k$$

where: $K = -(R + B'(Q+P)B)^{-1}B'(Q+P)A$

$$k = -(R + B'(Q+P)B)^{-1}(Q+P)\, \mathrm{E}[\varepsilon(\tau)] = 0$$

$$P = (A+BK)'Q(A+BK) + K'RK$$

and the prime denotes the transpose. Clearly the optimally controlled state now appears as

$$x(t+1) = (A+BK)x(t) + \varepsilon(t).$$

The matrix K is called the *feedback matrix* and the term k is called the *tracking gain* vector. Clearly if all the eigenvalues of the matrix $(A+BK)$ satisfy the condition of stability i.e. the absolute value of each eigenvalue is less than one, then the mean level of the optimal state would be stable. In case the parameters are time-varying, the feedback matrix and the tracking gain would be time-varying also e.g., $K(\tau)$, $k(\tau)$ and $u^*(\tau) = K(\tau)x(\tau\text{-}1)+k(\tau)$. The feedback matrix is usually backward looking, whereas the tracking gain is forward looking. When the state and control vectors are in deviation forms, these tracking gains $k(\tau)$ would contain nonzero terms involving $R(\tau)$, $Q(\tau)$ etc. and hence have been updated by reevaluating all $k(t)$ for $\tau \le t \le \mathrm{N}\text{-}1$, if any current expectation of future exogenous variable, e.g. $\mathrm{E}_\tau[e(\mathrm{N}\text{-}1)]$, is revised at τ or later. This sequential updating will be repeated at each point of the time horizon.

A second feature of the LQG model is the certainty equivalence property. This emphasizes the point that the optimal solution of the stochastic model is identical to the deterministic model which is obtained by replacing the vector of random variables $\varepsilon(t)$ by its expected value (i.e. zero in this case). Finally, the method of Kalman filters which underlies the LDR can be applied recursively so as to provide the sequential forecasts or updates of the conditional means and variances of $x(t)$. This aspect is discussed later.

We now consider a scalar dynamic model of linear decision rules, often arising in production-inventory models in microeconomics theory for illustrating the problem of jointness of estimation and control. Consider a simple discrete time control model with a fixed horizon: $x(t)$ is the state variable $u(t)$ is a control variable and t is time. The objective is to minimize

$$J = \tfrac{1}{2}\sum_{t=0}^{N-1}\left(x^2(t) + u^2(t)\right) \tag{4.2.7}$$

subject to

$$x(t+1) = \hat{a}y(t) + \hat{b}\,u(t) \tag{4.2.8}$$

where the initial condition $x(0) = x_0$ and the time horizon N are fixed. Here $\hat{a}$ and $\hat{b}$ are the ordinary least squares type estimates of the linear stochastic system

$$x(t+1) = ax(t) + bu(t) + \varepsilon(t) \tag{4.2.9}$$

with $\varepsilon(t) \sim N(0,\sigma^2)$, independent of $u(t)$ and serially uncorrelated, and where the parameters a, b are unknown constants to be estimated. A second method of estimation is to use (4.2.9) in (4.2.7) and minimize the expected value of J to arrive at the optimal equations for t=0,1,2,...,N-1

$$\begin{aligned} &x(t) + a\lambda(t+1) - \lambda(t) = 0 \\ &u(t) + b\lambda(t+1) = 0 \\ &x(t+1) - ax(t) - bu(t) = 0 \end{aligned} \tag{4.2.10}$$

where $\lambda(t+1)$ is the Lagrange multiplier associated with the state equation: $ax(t)+bu(t)+\varepsilon(t)-x(t+1) = 0$. On eliminating $\lambda(t)$ in (4.2.10) and a little algebraic manipulation, one can express optimal control $u^*(t)$ as a function of current and past states as follows:

$$\begin{aligned} &u^*(t) = \theta_1 x(t) + \theta_2 x(t-1), \\ &\theta_1 = (1-b^2)/(ab)\,, \text{ and } \theta_2 = -1/b. \end{aligned} \tag{4.2.11}$$

Alternatively, optimal control can also be expressed as

$$\begin{aligned} &u^*(t) = \beta_1 u(t-1) + \beta_2 x(t-1)\,, \\ &\beta_1 = (1+b^2)/a\,, \text{ and } \beta_2 = b \end{aligned} \tag{4.2.12}$$

With additive errors on the right-hand side of (4.2.11) or (4.2.12), one could again estimate the parameters β and θ, from which the structural parameter estimates $\hat{a}$ and $\hat{b}$ can be recovered. Note that this method, which has been widely used in empirical analysis of industry output behavior has several differences from the direct method used in (4.2.8). For one thing, it assumes that the observed data on the variable $u(t)$ are structured around the central value $u^*(t)$. Second, the estimates of the structural parameters $\hat{a}$ and $\hat{b}$ by these two methods are in general different. However, under certain situations the two estimates may be asymptotically equivalent. In control theory this is called the property of self-tuning Kumar and Varaiya (1986).

It is clear that by combining (4.2.9) and (4.2.12) one could define another method of estimation

$$\begin{pmatrix} u(t) \\ x(t) \end{pmatrix} = \begin{pmatrix} \beta_1 & \beta_2 \\ b & a \end{pmatrix} = \begin{pmatrix} u(t-1) \\ x(t-1) \end{pmatrix} + \begin{pmatrix} \varepsilon_1(t) \\ \varepsilon_2(t) \end{pmatrix}$$

where the asterisk has been dropped from $u(t)$. Note that this view of simultaneous estimation and control may be profitably exploited to build elements of caution into an optimal control policy based on the dynamic model (4.2.9). The difference in loss function (4.2.8) due to the two methods of estimation may be analyzed in terms of an iterative process involving estimation and control in successive steps. Thus one may estimate first the control equation (4.2.12) for $\hat{u}(t)$ given observations on $u(t\text{-}1)$ and $x(t\text{-}1)$. The second step consists in estimating $x(t+1)$ from (4.2.8) using the regressors $\hat{u}(t)$ and $\hat{x}(t)$. If the process converges in a few iterations, then this provide favorable evidence of robustness of either method of estimation. For econometric policy models, several applications made by Chow (1975), and Sengupta (1985) suggest that robustness may be the rule rather than an exception. This of course holds good, if there are no inequality restrictions on control or the state variables and the planning horizon is not too short.

4.2.2 Game theory Models.

We now consider dynamic models of differential games which arise in economic models of imperfect competition e.g. dynamic models of Cournot-type duopoly, the limit pricing model where a dominant supplier faces other suppliers and even models of conflict where the two or more suppliers compete for a fixed resource. Two such models are considered here, one due to Mangrel (1985) and the other to Roy et. al.(1994). These models exhibit two new characteristics. One is the specification of the dynamically optimal reaction curves for the two players, where the stability of the equilibrium solution depends on the parameters characterizing each player's reactions and risk attitudes. The second aspect is the estimation framework, where the likelihood function is maximized subject to the constraints imposed by the Cournot-Nash solutions. We discuss these two models briefly.

The first is a game-theoretic control model involving two countries. Each country i (i=1,2) must decide on the optimal amount of catch of fish c_i, which if uninterrupted, grows according to the biological law

$$x(t+1) = x^{\alpha}(t), \quad 0 < \alpha < 1. \tag{4.2.13}$$

It is assumed that this optimal decision is made by each country by maximizing the sum of the discounted utility of fish

$$V = \sum_{t=0}^{\infty} \beta^t u_i(c_i) = \sum_{t=0}^{\infty} \beta_i^t \log(c_i) \quad 0 < \beta_i < 1, \tag{4.2.14}$$

where β_i^t is the discounting function for each country and the utility $u_i(c_i)$ is taken in logarithmic form. If there is a one period horizon and if country 1 assumes country 2's action as given (e.g., zero conjectural variations in the classical Cournot model), its optimal control rule c_1 is easily found from the maximization problem

$$\max_{0 \le c_1 \le x - c_2} \{\log c_1 + \beta_1 \log[\tfrac{1}{2}(x-c_1-c_2)^{\alpha}]\}$$

i.e.

$$\max_{0 \le c_1 \le x - c_2} \{\log c_1 + \alpha\beta_1 \log(x-c_1-c_2) + \beta_1 \log \tfrac{1}{2}\} \tag{4.2.15}$$

where c_1 is the optimal catch of country 1, given the catch c_2 of country `2, whereas $(x-c_1-c_2)$ denotes the remaining stock of fish which grows to $(x-c_1-c_2)^{\alpha}$ in the next period. The first order condition for the maximum in (4.2.15) is

$$(1 + \alpha\beta_1)\, c_1 + c_2 = x \tag{4.2.16}$$

which specifies the reaction curve of country 1 in the one-period model. Similar arguments gives the optimal reaction curve of country 2 as

$$c_1 + (1 + \alpha\beta_2)\, c_2 = x \tag{4.2.17}$$

The equilibrium Cournot-Nash solution ($\bar{c}_1$, $\bar{c}_2$) given by the simultaneous solution of (4.2.16) and (4.2.17) is

$$\bar{c}_1 = D^{-1}\alpha\beta_2 x\,,$$

$$\bar{c}_2 = D^{-1}\alpha\beta_1 x\,, \tag{4.2.18}$$

$$y = x - \bar{c}_1 - \bar{c}_2 = D^{-1}\alpha^2\beta_1\beta_2 x$$

Next we consider a two-period horizon problem with country 1 still assuming that in the future, the one period Cournot-Nash solution (4.2.18) would prevail. The optimal value of the objective function under the one-period Cournot-Nash solution is

$$V_1(1) = \log \bar{c}_1 + \beta_1 \log\left(\tfrac{1}{2}\right) + \alpha\beta_1 \log\left(\bar{x} - \bar{c}_1 - \bar{c}_2\right) = (1 + \alpha\beta_1) \log (x) + A_1\,,$$

where

$$A_1 = \log\left(\frac{(\alpha\beta_2)(\alpha^2\beta_1\beta_2)^{\alpha\beta_1}}{D^{1+\alpha\beta_1}}\right) + \beta\log\left(\tfrac{1}{2}\right)$$

Since A_1 is a constant it will have no effect on the optimal policy for the two-year horizon. The objective function of country 1 for the two-year horizon is

$$J_1(2) = \log(c_1) + \alpha\beta_1(1+\alpha\beta_1)\log(x - c_1 - c_2) + A_1 .$$

Maximizing this function leads to the reaction curve of country 1

$$(1+\alpha\beta_1+\alpha^2\beta_1^2)c_1 + c_2 = x \tag{4.2.19}$$

Likewise the reaction curve of country 2 is

$$c_1 + (1+\alpha\beta_1+\alpha^2\beta_1^2)c_2 = x \tag{4.2.20}$$

This process can thus be repeated for an n-period horizon yielding the Cournot-Nash policies

$$\bar{c}_1 = D^{-1}(\alpha\beta_2)\left[\sum_{j=0}^{n-1}(\alpha\beta_2)j\right]x$$

$$\bar{c}_2 = D^{-1}(\alpha\beta_1)\left[\sum_{j=0}^{n-1}(\alpha\beta_1)j\right]x \tag{4.2.21}$$

where

$$D = \left[\sum_{j=0}^{n}(\alpha\beta_1)^j\right]\left[\sum_{j=0}^{n}(\alpha\beta_2)^j\right] - 1$$

and the remaining stock is:

$$y = x - \bar{c}_1 - \bar{c}_2 = D^{-1}\left\{\alpha^2\beta_1\beta_2\left[\sum_{j=0}^{n-1}(\alpha\beta_1)^j\right]\left[\sum_{j=0}^{n-1}(\alpha\beta_2)^j\right]x\right\}.$$

It is clear from the above that as $t \to \infty$, the limiting values of $\bar{c}_1$ and $\bar{c}_2$ become

$$\bar{c}_1 = D^{-1}(\alpha\beta_2)(1-\alpha\beta_1)x; \quad D = 1 - (1-\alpha\beta_1)(1-\alpha\beta_2) \tag{4.2.22}$$

$$\bar{c}_2 = D^{-1}(\alpha\beta_1)(1-\alpha\beta_2)x; \tag{4.2.23}$$

$$y = x - \bar{c}_1 - \bar{c}_2 = D^{-1}\alpha^2\beta_1\beta_2 x \tag{4.2.24}$$

To derive the steady-state level of $x(t)$ when the optimal steady state policies (4.2.22), (4.2.23) are pursued by the two countries, we consider the growth equation (4.2.13) under optimal policies for each country. Let $x_0 > 0$ be any initial stock of fish. Under Cournot-Nash equilibrium the growth of fish is governed by the difference equation

$$x(t+1) = [x(t) - c_1 x(t) - c_1 x(t)]^{\alpha} \tag{4.2.25}$$

At $t = 0$ we get

$$x(1) = (x(0) - c_1 x(0) - c_1 x(0))^{\alpha}$$

$$x(1) = x(0)[D^{-1}(\alpha^2\beta_1\beta_2)]^{\alpha},$$

where $D = 1 - (1-\alpha\beta_1)(1-\alpha\beta_2)$. At $t = 1$,

$$x(2) = [x(0)]^{\alpha^2}\left[D^{-1}(\alpha^2\beta_1\beta_2)\right]^{\alpha+\alpha^2},$$

and at $t = T$

$$x(T) = [x(0)]^{\alpha^T}\left[D^{-1}(\alpha^2\beta_1\beta_2)\right]^{\sum_{j=1}^{T}\alpha^j}$$

Hence as $T \to \infty$, $x(T) \to \bar{x} = D^{-1}(\alpha^2\beta_1\beta_2)^{\frac{\alpha}{1-\alpha}}$ If the discount rates are identical, then the steady state level of fish is:

$$\bar{x} = \left(\frac{\alpha\beta}{2-\alpha\beta}\right)^{\frac{\alpha}{1-\alpha}} \tag{4.2.26}$$

This solution may be compared with the cooperative solution when the two countries combine their catch so as to maximize the discounted sum of both countries' utilities, i.e.,

$$\max \sum_{t=0}^{\infty} \beta^t [2u(c(t))]$$

subject to

$$x(t) + 2c(t) = f(x(t-1))$$

$x(0) > 0$ given, $f(x) = x\alpha$, and $u(c) = \log c$.

The optimal solution in this case must satisfy the necessary condition

$$2\frac{\partial u(c(t))}{\partial c(t)} - 2\beta\frac{\partial u(c(t+1))}{\partial c(t+1)}\frac{\partial f}{\partial x(t)} = 0$$

from which one obtains the optimal growth equation for $x(t)$ as

$$x(t+1) = [\alpha\beta x(t)]\alpha \tag{4.2.27}$$

With an initial stock $x(0) > 0$ given, the steady state solution is

$$\hat{x} = \lim_{t\to\infty} x(t) = (\alpha\beta)^{\frac{\alpha}{1-\alpha}}. \tag{4.2.28}$$

On comparing (4.2.26) and (4.2.28) it is seen that $\hat{x} > \bar{x}$, i.e., the Cournot-Nash steady state equilibrium solution implies a smaller quantity of fish than the cooperative solution. Some generalized aspects of the solution structure may be noted here. First, other types of game-theoretic solutions, e.g., Stackelberg's leader-follower model, can be easily envisaged. Second, the linearity of the optimal decision rule in (4.2.21) or in (4.2.22) and (4.2.23) suggests, by the analogy of the feedback rule analyzed in the LQG model in Example 2, that if there are observational errors in the state variable x where $x = \bar{x} + \varepsilon$, ε being independent normal with zero mean and constant variance, one could still establish the approximate optimality of $\bar{c}_1$ and $\bar{c}_2$ in (4.2.22) and (4.2.23). This is so because x tends to $\bar{x}$ with probability one as the number of observations increases and we get

$$\bar{c}_1 \to k_1\bar{x}, \text{ and } \bar{c}_2 \to k_2\bar{x}$$

where

$$k_1 = D^{-1}(\alpha\beta_2)(1 - \alpha\beta_1), \text{ and } k_2 = D^{-1}(\alpha\beta_1)(1 - \alpha\beta_2).$$

This type of model has been empirically applied by Sengupta and Sfeir (1983) in the world coffee market, where it is assumed that Brazil is the first producer and the rest of the world represents the second producer. The solution of the two reaction curves then characterizes the Cournot-Nash equilibrium. Two sample periods were used: Sample I (1945-1962) and Sample II (1962-1980). The optimal reaction curves in terms of supply $x_1(t)$, $x_2(t)$ of Brazil and the rest of the world appear as follows:

Sample I:

$$x_1(t) = 5503 - .0385\, x_2(t) + .00001\, x_1^2(t)$$
$$x_2(t) = 71550 - 6.5001\, x_1(t) + .00005\, x_1^2(t)$$

Sample II: $x_1(t) = 33880 - .2985\, x_2(t) + .00001\, x_1^2(t)$
$x_2(t) = 56000 - .8401\, x_1(t) + .000005\, x_1^2(t)$

The results are significant: First, it is clear that in the second sample period, the influence of the rest of the world on Brazil has considerably (by about eight times). Secondly, the nonlinear reaction curves show that in the space of optimal supply, $x_1(t)$ and $x_2(t)$ are not always negatively correlated. This implies that at times Brazil may be able to improve the Cournot-Nash solution by exploiting the positive correlation of strategies e.g. in periods of frost damage in several countries, or world wide shortage in supply.

The second model considers a duopoly market with two rival brands, where the subscripts 1 and 2 are used to denote the sales quantities q^i_t and prices p^i_t, i=1,2. The sales equations at the beginning of period t are assumed to be linear:

$$q^1_t = a_{11}\, q^1_{t-1} + a_{12}\, q^2_{t-1} - b_{11}\, p^1_t + b_{12}\, p^2_t + u^1_t$$

$$q^2_t = a_{21}\, q^1_{t-1} + a_{22}\, q^2_{t-1} + b_{21}\, p^1_t - b_{22}\, p^2_t + u^2_t\,.$$

Further it is assumed that each brand sets prices so as to minimize the discounted sum of squared deviations from the target vectors $(q^1)^o$, $(q^2)^o$ over T periods. On using the same discount rates r for the two players, one obtains the objective function:

$$\underset{p_1,p_2\cdots p_T}{\text{Minimize}}\ J = E_t\left[\sum_{t=1}^{T} r^{t-1}\left\{\left(q^1_t - \left(q^{11}\right)^0\right)^2 + \left(q^2_t - \left(q^{12}\right)^0\right)^2\right\}\right]$$

where the target vectors are $(q^1)^o = ((q^{11})^o, (q^{12})^o)$, and $(q^2)^o = ((q^{21})^o, (q^{22})^o)$. Assuming the error terms u^1_t and u^2_t to be independently distributed as Gaussian variables with zero means and constant variance, this model is in the LQG framework, since the objective function is quadratic. Hence the optimal pricing rules for brands 1 and 2 can be derived as linear decision rules:

$$p^{1*}_t = g^1 + G^1 q_{t-1} + S^1(f^1_t - \bar{f}^{\,1}) \quad \text{and} \quad p^{2*}_t = g^2 + G^2 q_{t-1} + S^2(f^2_t - \bar{f}^{\,2}),$$

where g^1 is the scalar intercept of brand one's price rule, $G^1 = [G^{11}, G^{12}]$ is a vector whose first elements reflect the weight given brand one's own lagged sales and the second is the weight attached to brand two's lagged sales, $S^1 = [S^{11}, S^{12}]$ is a vector whose first elements reflect the weight attached to the deviation of the forecast from the mean forecast prior to period t for brand one, and the second element is for the other brand, $q_{t-1} = [q^1_{t-1}, q^2_{t-1}]'$ is the vector of lagged sales, $f^1_t = [f^{11}_t, f^{12}_t]'$ is the vector of forecasts for brand one, $\bar{f}^{\,1}_t = [\bar{f}^{\,11}_t, \bar{f}^{\,12}_t]'$ is the average of forecasts made by one, and up to time t-1. With no discounting, the parameters of the optimal price equation appears as follows:

$$g' = (B_1'B_1)^{-1}B_1'[(q^1)^o - B_2g^2]$$

$$G' = (B_1'B_1)^{-1}B_1'(A+B_2G^2]$$

$$S' = -G'(A+G'^1B'_1+G'^2B'_2)\Sigma\Gamma_1^{-1}$$

where

$$A = \begin{bmatrix} a_{11} & a_{12} \\ a_{21} & a_{22} \end{bmatrix}, \qquad B_1 = (b_{11}, b_{21})', \qquad B_2 = (b_{12}, b_{22})',$$

Σ denotes the sales covariance matrix for the error terms u_t^1, u_t^2 and Γ_1 is the covariance matrix between forecasts made by brand one, and actual sales. Note that a similar set of variables and equations can be derived for the other brand. The above sets of optimal pricing rules can be empirically estimated by using maximum likelihood equations, since one can specify the likelihood function for the sales equations and the complete set of first order conditions given by the LDR. Hence the loglikelihood function can be maximized subject to the two sets of first order conditions specifying the LDR, one for each brand. Gasmi, Laffont and Vuong (1990) have shown that this maximum likelihood estimation procedure will yield statistically consistent estimates of Cournot-Nash equilibrium behavior with a leader and a follower.

4.2.3 A Dynamic Portfolio Model.

Finally we consider a dynamic model of optimal savings and investment under logarithm preferences. This model considers an agent who solves for the optimal consumption profile $\{c_t\}$:

$$\max E_0\left[\sum_{t=0}^{\infty} \beta^t \ln c_t\right], \quad 0 < \beta < 1$$

such that $w_{t+1} = r_t(w_t - c_t)$, $t \geq 0$, $w_0 > 0$ is given, β is a discount factor, and where E_0 is expectation as of time zero, w_t is the wealth or assets at the beginning of period t and the only source of stochastic uncertainty is the random r_t, which is governed by a first order Markov process and becomes known at the beginning of period t+1. The agent also faces the standard constraint that infinite consumption financed through infinite borrowing is not feasible.

Form the Lagrangean function

$$L = E_o\left[\sum_{t=0}^{\infty}\beta^t\left\{\ln c_t + \lambda_{t+1}\left(r_t w_t - r_t c_t - w_{t+1}\right)\right\}\right]$$

and apply the first order necessary conditions for optimality

$$\frac{\partial L}{\partial w_t} = 0 = \beta^t \mathrm{E}_t[r_t \lambda_{t+1}] - \beta^{t-1}\lambda_t$$

$$\frac{\partial L}{\partial c_t} = 0 = \beta^t \mathrm{E}_t[c_t^{-1} - r_t \lambda_{t+1}] \tag{4.2.29}$$

$$\frac{\partial L}{\partial \lambda_{t+1}} = 0 = r_t w_t - r_t c_t + w_{t+1} \ .$$

By the strict concavity of the utility function $u(c) = \ln c_t$ these necessary conditions are sufficient for the maximum. The system (4.2.29) yields after some algebraic manipulations the following two important relations for the optimal consumption profile $\{c_t^*\}$

$$\frac{1}{c_t^*} = \beta E_t\left[\frac{r_t}{c_{t+1}^*}\right] \tag{4.2.30}$$

$$c_t^* = (1-\beta)w_t^* = (1-\beta)\beta^t\left(\prod_{s=0}^{t-1} r_s\right)w_0^* \tag{4.2.31}$$

$$c_0^* = (1 - \beta)w_0^*$$

The first relation (4.2.30) is easily derived from the first two equations of (4.2.29) by a process of substitution. The second relation (4.2.31) may be obtained by using the Riccati transformation $\lambda_t^* = p_t/w_t^*$ and deriving the linear difference equation

$$p_{t+1} - \beta^{-1}p_t + 1 = 0$$

which has the solution

$$p_t = A\,\beta^{-t} + (1 - \beta)^{-1}\beta$$

But since β^{-1} exceeds unity, p_t goes to infinity as t becomes infinite. Hence for a stable consumption profile one has to set the constant A to zero. This yields $p_t = (1 - \beta)^{-1}\beta$ and hence

$$\frac{1}{\lambda_t^*} = \frac{w_t^*}{\beta/(1-\beta)} = \beta^{-1}(1-\beta)w_t^*$$

$$c_t^* = \beta / \lambda_t^* = (1-\beta)\, w_t^*$$

Several interesting implications follow from these optimal portfolios of consumption and saving. First of all, optimal consumption c_t^* at time t depends only on past returns up to t-1 and it does not depend on any variable which is not part of the agent's information set at time t. The main reason for this strong result is the logarithmic utility function. For example, if we have an utility function with a constant relative risk aversion

$$u(c_t) = (1-\rho)^{-1} c_t^{1-\rho}, \quad \rho > 0\,,$$

and assume that r_t is independently and identically distributed with $E[r_t^{1-\rho}] < \beta^{-1}$, then the optimal consumption program becomes

$$c_t^* = [1 - \beta^{1/\rho}\{E[r_t^{1-\rho}]\}^{1/\rho}]w_t^*$$

Clearly c_t^* depends on the return process $\{r_t\}$. Secondly, the model can be generalized to include many assets in the portfolio by redefining returns as

$$r_t = \sum_{j=1}^{n} \alpha_{jt}\, r_{jt}\,, \quad \sum_{j=1}^{n} \alpha_{jt} = 1 \quad \alpha_{jt} = 0$$

where $\alpha_t = (\alpha_{jt})$ is the portfolio composition of n assets. Given α_t, the conditional distribution of r_t can be utilized in defining the optimal consumption program in (4.2.30) and hence the optimal value of the objective function in (4.2.28). Let $J^*(\hat{\alpha}_t)$ be the optimal value of the objective function for a particular portfolio. Assume that the agent is choosing between two portfolios e.g. one with an equal size portfolio i.e., $\alpha_{jt} = 1/n$ for all j and the other a minimum variance portfolio which minimize the variance of r_t subject to a lower bound on expected return. The final optimal choice is then given by the minimum of $J^*(\hat{\alpha}_t^{(1)})$ and $J^*(\hat{\alpha}_t^{(2)})$. Thus the minimum of $J^*(\hat{\alpha}_t^{(k)})$ over a set $k = (1,2,...,K)$ of choices decides the final policy. Thirdly, one could compare the optimal stochastic program (4.2.30) with the deterministic control, where the expectation term is dropped in (4.2.31). Generally however, the deterministic set up would fail to generate a very accurate prediction of the level of consumption or of wealth. Finally, different forms of the probability distribution of the random variable r_t would influence the optimal consumption program. Thus if the $\{r_t\}$ process has significant departures from a Gaussian process, the skewness and kurtosis of the optimal consumption program would have major implications for the agent's decision process.

4.3 Kalman filtering methods

Methods of Kalman filtering provide some useful ways of building adaptivity into the estimation process. With more information the estimates can be updated and hence a recursive procedure can be easily developed. Consider a simple example.

Let the system dynamic with a scalar input u_t and a scalar output x_t be given as

$$x_t + ax_{t-1} = u_{t-1} + e_t + ce_{t-1} \tag{4.3.1}$$

where it is assumed that for all time points $t = 1,2,\ldots,$ the stochastic process $\{e_t\}$~NID(0,1) is normally independently distributed with zero mean and unit variance and the parameters a, c where $|c| < 1$ are known. The problem is to choose a control law $\{u_t\}$ which minimizes the variance of output, $E[y_t^2]$ at each t in the future. Alternatively, the problem may be stated as a case of pure prediction of forecasting for the system

$$x_t + ax_{t-1} = e_t + ce_{t-1} \tag{4.3.2}$$

Consider the time point t when the set X_t:$\{x_t, x_{t-1}, \ldots x_1, x_0\}$ of observation on the state x_t is available. We have to forecast or estimate a value $\hat{x}$ of x_{t+1} at the future time point $(t+1)$, such that $\hat{x}$ is a linear function of the available observation set X_t and it has minimum variance or prediction error. Whereas the first specifies a control-theoretic view, linear quadratic Gaussian (LQG) to be exact, the second describes the linear least squares (LS) view of the Kalman filtering techniques. The duality in interpretation is sometimes called the separation principles, which is more generally applicable. For example, inputs and outputs may be vectors, time may be continuous and even the Gaussian assumption may be somewhat relaxed.

Note that by setting $\varepsilon_t = e_t + ce_{t-1}$ one would get back a dynamic model very similar to the one (4.3.1) considered before, except that the control variable has a lagged impact on x_t (i.e., control today, u_t affects output tomorrow x_{t+1}). This type of model is called casual, in the sense that the cause variable (u_t) precedes the effect variable (x_{t+1}). Hence the conditional argument of linear LS theory is applicable e.g.

$$E[x_t|x_{t-1},u_{t-1}] = -ax_{t-1} + u_{t-1}$$

In econometric theory, x_{t-1} and u_{t-1} are called predetermined rather than exogenous variables. Hence the ideal conditions of linear LS theory, which require x_{t-1} and u_{t-1} to be truly exogenous and independent of $\varepsilon_t = e_t + ce_{t-1}$, and ε_t to satisfy

$\{e_t\}$~NID(0,1), are not exactly fulfilled. However, methods like Cochrane-Orcutt procedure discussed in standard econometric texts may be adopted to modify the LS procedure slightly in order to obtain unbiased and consistent estimates of the relevant parameters at least asymptotically.

For the minimum variance control problem (4.3.1), it is clear that

$$E[x^2_{t+1}] \geq E[e^2_{t+1}] = 1.$$

Since e_{t+1} is assumed to be independent of x_t, u_t and e_t. Hence the best control law can only make the variance of output greater than or equal to unity. It is clear that if we choose

$$u_t = ax_t - ce_t \tag{4.3.3}$$

then $x_{t+1} = e_{t+1}$ which gives the smallest possible variance of x_{t+1}. If this control law is used in every time point, one would get $u_t = ax_t - cx_t = (a-c)x_t$, since $e_t = x_t$ under this control law. Hence the dynamic equation (4.3.1) becomes

$$x_{t+1} + cx_t = e_{t+1} + ce_t,$$

or

$$x_{t+1} - e_{t+1} = - c(x_t - x_t),$$

Solving this difference equation with initial value $x_{t_0} - e_{t_0} = \tau_0$ we get

$$x_t = e_t + \tau_0(-c)^{(t - to)},$$

and since $|c| < 1$,

$$\lim_{t_0 \to -\infty} (x_t) = e_t \ . \tag{4.3.4}$$

Hence, if the above control law (4.3.3) is used, the output in the steady state has the smallest possible variance of unity.

Some interpretations of this minimum variance control law are in order. The quantity $\hat{x} = -ax_t + bu_t + ce_t$, where of course $b = 1$ in this simple example can be interpreted as the best estimates in the sense of minimum MSE of output at time t+1 based on the data available up to time t. The prediction error is e_{t+1}. By choosing the optimal control law according to (11.3) we thus make the predicted value of x_{t+1} equal to the desired value of zero. Second, one could obtain the same result, if the normality assumption on the e_t is dropped but the admissible control laws are still restricted to be linear functions of available observations. We should then assume that e_t has zero mean and unit variance for all t, that e_t and e_s are uncorrelated for all $t \neq s$, and that u_t is a linear function of the information set X_t:$\{x_t, x_{t-1}, \dots, x_0\}$. Third, if one interprets x_t as deviations from some fixed targets

x^* say, then we have a "*tracking problem*": the optimal control rule minimizes a measure of closeness to the target.

Now consider the alternative specifications of the linear dynamic (4.3.3), which in operator notation $D^j x_t = x_{t-j}$ becomes

$$x_{t-1} = \frac{1+cD}{1+aD} e_{t+1} = e_{t+1} + \frac{c-a}{1+aD^{e_t}} \tag{4.3.5a}$$

but

$$e_t = \frac{1+aD}{1+cD} x_t \tag{4.3.5b}$$

On eliminating e_t between (4.3.5a) and (4.3.5b) one gets:

$$x_{t+1} = e_{t+1} + \frac{c-a}{1+cD} x_t \tag{4.3.5c}$$

Now let $\hat{x} = \hat{x}_{t+1|X_t}$ be any prediction of x_{t+1} as a linear function of the available observations contained in the set X_t defined above. It is assumed that the future error, e_{t+1} is statistically independent of the observations $(x_t, x_{t-1}, \ldots, x_1, x_0)$ i.e.,

$$e_{t+1} \perp X_t, \quad \text{for all } t. \tag{4.3.6}$$

Hence we have: $E\left[x_{t+1}\ \hat{x}_{t+1|X_t}\right]^2 = E\left[e_{t+1}^2\right] + E\left[\frac{c-a}{1+cD} x_t - \hat{x}_{t+1|X_t}\right]^2.$

Therefore $E\left[x_{t+1} - \hat{x}_{t+1|X_t}\right]^2 \geq E\left[e_{t+1}^2\right] = 1$, where the inequality reduces to equality, if the forecast value $\hat{x}$ satisfies: $\hat{x} = \hat{x}_{t+1|X_t} = \frac{c-a}{1+cD} x_t$, which is nothing but the difference equation

$$\hat{x}_{t+1|X_t} + c\hat{x}_{1|X_{t-1}} - (c-a)x_t = 0 \tag{4.3.7}$$

Under this optimal forecast (4.3.7), the minimum value of the MSE($\hat{x}$) is

$$MSE\left(\hat{x}_{t+1|X_t}\right) = E\left[x_{t+1} - \hat{x}_{t+1|X_t}\right]^2 = E\left[e_{t+1}^2\right] = 1.$$

This result is identical with that of (4.3.3) and (4.3.4).

Now consider a two-step predictor which is optimal in the sense of minimum MSE of forecast. In this case the observation set X_t is as before: we are at point t. We want to predict the future value x_{t+2} by a two-step predictor, denoted by $\hat{x}_{t+2|X_t}$, which, as before must be a linear function of the available observations in the set X_t, and minimize the $MSE(\hat{x}_{t+2|X_t}) = E[x_{t+2} - \hat{x}_{t+2|X_t}]^2$. By following procedures analogous to (4.3.5), we obtain:

$$x_{t+2} = e_{t+2} + (c-a)e_{t+1} - \frac{a(c-a)}{1+cD}x_t \tag{4.3.8}$$

By the assumption (4.3.6), the error terms e_{t+1} and e_{t+2} are independent of the observations in the set X_t, hence we have:

$$MSE(\hat{x}_{t+2|X_t}) = E[x_{t+2} - \hat{x}_{t+2|X_t}]^2 = E[e_{t+2}^2] + (c-a)^2 E[e_{t+1}^2] + E\left[\hat{x}_{t+2|X_t} + \frac{a(c-a)}{1+cD}x(t)\right]^2$$

therefore $MSE(\hat{x}_{t+2|X_t}) \geq 1 + (c-a)^2$, where the equality sign hold for the optimal forecast $\hat{x}_{t+2|X_t}$ which must satisfy $\hat{x}_{t+2|X_t} = -\frac{a(c-a)}{1+cD}x(t)$. The two-step optimal predictor thus satisfies the difference equation

$$\hat{x}_{t+2|X_t} + c\hat{x}_{t+1|X_{t-1}} + a(c-a)x_t = 0.$$

If $\eta_{t+2|t}$ denotes the error of the two-step optimal predictor i.e., $\eta_{t+2|t} = x_{t+2|t} - \hat{x}_{t+2|X_t}$, then it follows from (4.3.8) that: $\eta_{t+2|t} = e_{t+2} + (c - a)e_{t+1}$ (i.e., the two-step optimal prediction error is a moving average of second order). In this notation, the one-step optimal predictor is a moving average of first order, since from (4.3.5c) and (4.3.7), $\eta_{t+1|t} = e_{t+1}$.

A more general result giving a k-step ahead optimal predictor $\hat{x}_{t+k|X_t}$, where the latter is a linear function of the available observations X_t and minimizes the MSE: $E[x_{t+k} - \hat{x}_{t+k|X_t}]^2$ is available in Box and Jenkins (1976), and Kumar and Varaiya (1986). Two points about the k-step optimal predictor ($k>1$) are worth noting. First, it is a moving average of order k, hence the averaging process of future errors $e_{t+1}, e_{t+2}, \ldots, e_{t+k}$, if correctly done is bound to improve the forecast. This updating process may not always be exactly realized in real life situations, since the parameters a, c and σ_e^2, which are assumed to be known here, are rarely known. Besides, if the inputs and outputs are vector variables and not scalars, the updating process may involve substantial computations. This will be shown in the next section. Second, the most critical assumption in all these optimal k-step forecasts is that they are independent of future errors $e_{t+1}, e_{t+2}, \ldots, e_{t+k}$ mutually and from the observations $x_t, x_{t-1}, \ldots, x_1, x_0$. Due to this assumption one gets $E[\eta_{t+k|t}] = 0$ and $E[\eta_{t+k|t}\ \eta_{t+j|t}'] = 0$ for $k \neq j$ (i.e., the successive one-step ahead optimal forecast errors are unbiased and uncorrelated). Hence under the given conditions, no better forecasts can be constructed.

Now consider the system dynamics (4.3.1) in a vector-matrix form, where x_t and u_t are now n-element and m-element column vectors, A, B and C are matrices assumed for simplicity to be time-independent and v_t is measurement noise for outputs x_t:

$$x_{t+1} = Ax_t + Bu_t + e_{t+1} \tag{4.3.9}$$

$$y_t = Cx_t + v_t \tag{4.3.10}$$

where $E[e_t] = 0$ for all t, independent over time, $E[e_t e_t'] = R_1$, $E[v_t] = 0$ for all t, independent over time and independent of e_t, and $E[v_t v_t'] = R_2$ where R_1 and R_2 are constant positive definite matrices. Also, the initial value x_0 at $t = 0$ is assumed for simplicity to be normally and independently distributed with known mean m and covariance matrix R_0. The problem again is to make an optimal predictions $\hat{x}_{t+1}$ of x_{t+1} given the observations $\{y_t, y_{t-1}, \ldots, y_0\}$, which of course has additive noise components. Note that this model is *causal* in the sense that the control vector u_t precedes x_{t+1}; however x_t is not directly observable except through noisy measurements by vector y_t. Thus, economic models with permanent and transitory components fit this specification.

Let the prediction of x_{t+1} be constructed by the following linear decision rule (LDR)

$$\hat{x}_{t+1} = A\hat{x}_t + Bu_t + K_t(y_t - C\hat{x}_t) \tag{4.3.11}$$

where K_t is a matrix to be determined below. Denote $(x - \hat{x})$ by $\tilde{x}$, and then from (4.3.9) through (4.3.11) one could derive

$$\tilde{x}_{t+1} = (A - K_tC)\tilde{x}_t + e_{t+1} - K_tv_t \tag{4.3.12}$$

which has the mean $E[\tilde{x}_{t+1}] = (A - K_tC)E[\tilde{x}_t]$, which converges to zero if the initial condition is set such that $\hat{x}_0 = m$ and the matrix $(A - K_tC)$ has all its eigenvalues less than unity in absolute value. Let V_{t+1} be the variance-covariance matrix of $\tilde{x}_{t+1}$. Then it follows from (4.3.12) that V_{t+1} satisfies the linear difference equation

$$V_{t+1} = (A - K_tC)V_t(A - K_tC)' + R_1 + K_tR_2K_t' \ , \tag{4.3.13}$$

with the initial condition

$$V_0 = R_0 \, . \tag{4.3.14}$$

Now we choose the matrix K_t (the Kalman gain) by a MSE criterion. That K_t is chosen which minimizes the variance of the scalar product $w'\tilde{x}_{t+1}$ (i.e. $w'V_{t+1}w$), where the vector w of nonzero weights is arbitrary. Since the variance of prediction error is now a matrix, this scalar version of the MSE criterion gives determinate results. On using (4.3.14) in the scalar MSE criterion $x'V_{t+1}w$ and minimizing with respect to K_t gives the optimal value of K_t as

$$K_t = AV_tC'(R_2 + CV_tC'\}^{-1} \tag{4.3.15a}$$

when

$$V_{t+1} = AV_tA' + R_1 - (AV_tC')[R_2 + CV_tC'](CV_tA') . \tag{4.3.15b}$$

Several features of this optimal gain matrix K_t are to be noted. First, it is independent of the weights w used to define the scalar MSE criterion. Hence the result would hold even if we use a different set of weights. Second, the solution for K_t involves the variance covariance matrix V_t, which has to be solved from (4.3.15b), which is a set of quadratic matrix equations (the Riccati equations). The equations (4.3.15) can only be solved recursively through an iterative process and this may impose great computational burden, if the size of the square matrix V_t which is of order n is large. Moreover, the requirement from (4.3.12) that $E[\hat{x}_{t+1}]$ converges to zero as t becomes infinite depends on the fact that the eigenvalues of the matrix $(A\text{-}K_tC)$ are less than unity i.e., this ensures asymptotic stability of the mean. Thus, for every t, K_t should be positive definite and in the limit, K_t must be finite as t becomes infinite. If K_t becomes singular or, nearly so there arise additional computational difficulties. Note that if the system is asymptotically stable and $\bar{V}$ denotes the steady state value of V_t, then $\bar{V}$ can be solved from the algebraic matrix equation:

$$\bar{V} = A\bar{V}A' + R_1 - (A\bar{V}C')[R_2 + C\bar{V}C]^{-1}(C\bar{V}A)$$

and the steady state value of K_t becomes

$$\bar{K} = \left(A\bar{V}C'\right)\left(R_2 - C\bar{V}C'\right)^{-1} .$$

Using this approximation, the updating linear decision rule (4.3.11) becomes

$$\hat{x}_{t+1} = A\hat{x}_t + Bu_t + \bar{K}\left(y_t - C\hat{x}_t\right) \tag{4.3.16}$$

which shows that the grain matrix $\bar{K}$ allows an adjustment of the first stage forecast denoted by $(A\hat{x}_t + Bu_t)$ by exploiting the difference $(y_t - C\hat{x}_t)$ between the actual measurements and their predictions. Note further that since we view Kalman filter as a pure prediction problem, the coefficient B or u_t do not appear anywhere in the optimal equations (4.3.15a) and (4.3.15b). Note however that once the optimal K_t is known, the optimal control u_t can be chosen as a function of $\hat{x}_t$. In capital stock adjustment models, the flexible accelerator principle showing investment to be proportional to the difference between actual and desired capital stock has been frequently used in this form (4.3.16). The Kalman gain helps to provide an optimal interpretation of the distributed lags in the adjustment cost theory, which has been used most frequently in economic models e.g. aggregate production-inventory models, monetary-fiscal policies under a macrodynamic model and commodity stabilization policies. Some other important economic applications include price adjustment in a Marshallian market, exponentially weighted forecasting rules and various types of adaptive control policies in commodity markets.

One more point about the variance equation (4.3.15b) which must be recursively solved for computing optimal Kalman gain matrices K_t in (4.3.15a). Note that this variance, which is a matrix measure of the errors of prediction is a sum of three terms: the first i.e. AV_tA' shows how the error propagates to stage $(t+1)$ through system dynamics (i.e. if A is not known exactly, this would contribute to additional noise), the second term R_1 shows the increase of error variance due to the disturbances e_t and the third term shows how the measurement noise in output affects the variance-covariance matrix V_{t+1} through the matrix R_2.

4.3.1 Applications of Kalman Filter

We consider in this section two applications of the Kalman filter, which are important from an economic viewpoint. The first involves near-optimum systems in linear-quadratic control problems, when the objective is to achieve a prescribed degree of stability. Since there is a trade-off problem between optimality in the sense of fine tuning and stability in the sense of obtaining a steady state goal of the controlled state, it is desirable to study the problem of obtaining a prescribed degree of stability. The second model analyses the dynamics of the mean-variance efficiency frontier in an intertemporal investment model, where the various problems of estimation and sequential updating are discussed.

A. Control with a prescribed degree of stability

Consider first a discrete linear control system where $x(k)$ is a state vector of dimension n, and $u(k)$ is a control vector of dimension m. The optimal control model is:

$$\text{Min } J = \tfrac{1}{2}\sum_{k=0}^{\infty}\left[\alpha^{2k}\left(x'(k)Qx(k)+u'(k)Ru(k)\right)\right] \tag{4.3.17}$$

subject to $x(k+1) = (I+A)x(k) + Bu(k)$, where Q and R are positive-semidefinite and positive definite matrices respectively, and $\alpha > 1$. Since α^{2k} can be interpreted as a discount function, this type of objective function may be interpreted as a risk adjusted performance function. Define the following transformations: $\hat{x}(k) = \alpha^k x(k)$, and $\hat{u}(k) = \alpha^k u(k)$. The model can now be written as

$$\text{Min } J = \tfrac{1}{2}\sum_{k=0}^{\infty}\left[\left(\hat{x}'(k)Q\hat{x}(k)+\hat{u}'(k)R\hat{u}(k)\right)\right] \tag{4.3.18}$$

subject to $\hat{x}(k+1) = \left(\hat{A}\hat{x}(k)+\hat{B}\hat{u}(k)\right)$, where $\hat{A} = (\alpha I + \alpha A)$, and $\hat{B} = \alpha B_1$. For the system (4.3.18), the optimal control is given by the linear feedback rule:

$$\hat{u}(k) = -\left(R+\hat{B}'P\hat{B}\right)^{-1}\hat{B}'P\hat{A}\hat{x}(k) = -K\hat{x}(k).$$

This implies that $u(k) = -Kx(k)$, where P is the unique positive semi-definite solution of the Riccati equation: $P = \hat{A}'P\hat{A}+Q-\hat{A}'P\hat{B}\left(\hat{B}'P\hat{B}+R\right)^{-1}\hat{B}'P\hat{A}$. The closed loop form of the controlled system is given by:

$$x(k+1) = [A - Bk]x(k) \tag{4.3.19}$$

whereas for the transformed system (4.3.18) one obtains

$$\hat{x}(k+1) = \alpha[A - Bk]\,\hat{x}(k) \tag{4.3.20}$$

For asymptotic stability we must have all the eigenvalues of $[A - Bk]$ to lie inside the unit circle in the complex plane for the system (4.3.19), whereas for (4.3.20) the eigenvalues of $\alpha[A - Bk]$ must be inside the circle of radius $1/\alpha$ in the complex plane. Thus α turns out to be the prescribed degree of stability. By appropriately choosing α, the convergence to a steady state may be made faster. This also shows the importance of weighting in systems which tend to be unstable.

B. A Dynamic view of portfolio efficiency

The specification and estimation of the mean-variance efficiency frontier in portfolio theory have generally followed three approaches. One minimizes the variance of portfolio returns subject to a lower bound on the expected return. This leads to the efficiency frontier, where the mean vector m and the covariance matrix V of returns of individual stocks in the portfolio have to be estimated. The second approach uses the capital asset pricing model to specify the frontier as the linear capital market line, where the conditional expected return of a portfolio (or, mutual fund), net of the risk free-rate of return is viewed as a linear function of the market return, net of the risk-free rate. This approach which leads to the estimates of the regression coefficients otherwise known as the β coefficients involves only static aspects of the frontier, although statistical tests are usually made in empirical work to check if the coefficients tend to vary over time. The third approach considers the class of multiperiod portfolio policies in terms of a dynamic horizon for the investor. The specification of a dynamically optimal portfolio policy in this framework offers some interesting generalizations to the one-period model of portfolio management e.g., it shows that the optimality of a suitable multiperiod policy may imply a myopic policy in suitable situations, and secondly, it may suggest how to build robustness aspects into the portfolio decision rules when myopic policies are not optimal. Our object in this section is to explore some stochastic aspects of the dynamically optimal class of multiperiod portfolio policies of a representative investor. Specifically we consider the stochastic control aspects of such optimal portfolio policies, when some of the parameters have to be estimated or, the estimates have to be updated. Robustness aspects arise very naturally in this situation through the presence of estimation risk and noisy information.

The model of intertemporal efficiency frontier we consider is adapted from the formulations of Bertsekas (1970). There exist several alternative criteria of optimization in the current literature for the specification of the dynamic portfolio frontier e.g., the expected discounted final wealth with a stochastic discount factor, or the expected sum of discounted future utility of consumption, given a budget constraint. Here for simplicity we assume a utility based asset pricing model which can be easily related to the mean-variance approach. We assume the investor to be maximizing the expected value of his final wealth $E[U(W_T)]$, where the wealth is defined by a discrete-time control system:

$$W_{t+1} = s_t\left(W_t - \sum_{i=1}^{n} u_{i,t}\right) + \sum_{i=1}^{n} b_{i,t}u_{i,t} = s_t W + \sum_{i=1}^{n} r_{i,t}u_{i,t}\,, \quad t = 0,1,\ldots,T\text{-}1 \quad (4.3.21)$$

where s_t, $b_{i,t}$ are the rates of return of the riskless asset and the i-th risky asset in period t respectively with $r_{i,t} = b_{i,t} - s_t$ and $u_{i,t}$ is the amount invested at the beginning of period t in the i-th risky asset. We assume s_t to be non random and the vector $\boldsymbol{r}_t = r_{i,t}$ to be random with mean $\boldsymbol{m}_t$ and covariance matrix V_t. Since the data on random returns may involve noises e.g., $\boldsymbol{r}_t = \boldsymbol{m}_t + \varepsilon_t$, and also in some cases (e.g., short horizons) s_t and $\boldsymbol{r}_t$ may not vary over time, we may characterize the above system dynamics in several versions as follows:

$$\begin{aligned} W_{t+1} &= sW_t + r'\boldsymbol{u}_t \\ W_{t+1} &= sW_t + \boldsymbol{m}'\boldsymbol{u}_t + \varepsilon'\mathbf{u}_t \\ W_{t+1} &= sW_t + \boldsymbol{m}'\boldsymbol{u}_t + \zeta_t \qquad (4.3.21a) \\ W_{t+1} &= sW_t + \tilde{r}_t\,\boldsymbol{u}_t + \zeta_t \end{aligned}$$

where we may assume for simplicity that the random component ζ_t is identically and independently distributed with mean zero and variance V_ζ and likewise for the noisy components ε_t having zero means and a covariance matrix V_ε. As for the utility function $U(W_T)$ of the terminal wealth we have two distinct cases to consider:

Case I: $U(W) =$ either $-e^{-W/a}$ or $\ln(W + a)$ (4.3.22)

Case II: $U(W)$ does not satisfy (4.3.22), i.e., the risk tolerance function

$$-\partial U(W_t)/\partial^2 U(W_t) \neq a + bW_t, \qquad \text{for all } t = 0,1, \ldots T. \tag{4.3.23}$$

As Mossin (1968) has shown that if the probability distribution of the yield vector $\boldsymbol{r}_t$ is identical in all periods and the utility function satisfies the condition (4.3.22) of linearity of the risk tolerance function, then the investor's optimal portfolio policy is "stationary" or "myopic" in the sense that the same proportion is invested in each asset in every period, i.e., the investor when faced with the opportunity to reinvest sequentially his wealth can use a policy similar to the single period case. However this type of myopic policy is not optimal under Case I if the yield vector r_t is not stationary over time, or if the utility function does not satisfy the condition (4.3.22). We have to analyze these cases separately. Consider first the case of a quadratic utility function $U(W) = W - kW^2$, where to keep conformity with the Markowitz-Tobin mean-variance model we formulate the investor's optimizing model as follows:

$$\text{Min } \mathrm{E}[W_T^2]$$

Subject to:

$$\mathrm{E}[W_T] \leq c_T \qquad W_{t+1} = sW_t + r'\boldsymbol{u}_t; \quad \boldsymbol{u}_t \in A, \quad 0 \leq t \leq T\text{-}1.$$

Here A is a suitable convex set of the class of controls $\boldsymbol{u}_t$ and the nonrandom quantity c_T may be related to the market portfolio as in the capital asset pricing theory. Since

$$\mathrm{E}W^2_t = \mathrm{E}\{\mathrm{E}[W^2_t \mid W_0, W_1, \ldots, W_{T-1}, u_0, u_1, \ldots, u_{T-1}]\}$$

where the outer expectation operation is taken with respect to the random variables $W_0, W_1, \ldots, W_{T-1}$ which depends on r_t's. Clearly $\mathrm{E}W^2_t$ is minimized by minimizing the inner conditional expectation with respect to u_{T-1} for every possible collection of $(W_0, W_1, \ldots, W_{T-1})$. This leads to the optimal linear decision rule (LDR):

$$u^*_{T-1} = R^{-1}\boldsymbol{m}(c_T - sW_{T-1})/\alpha,$$
$$R = \mathrm{E}[\boldsymbol{rr}'], \quad \alpha = \boldsymbol{m}'R^{-1}\boldsymbol{m} \tag{4.3.24}$$

and the associated minimal variance of W_t as

$$\sigma^2_T{}^* = (c_T - sW_{T-1})^2(1-\alpha)/\alpha, \tag{4.3.25}$$

where $W_{T-1} = h_{T-1}(W_0)$ an be specified as a suitable nonlinear function of the initial value W_0 by the dynamic programming algorithm. The LDR (5.1) may be compared with two other plausible decision rules. One assumes the certainty equivalence principle and replaces the system dynamics as

$$W_T = sW_{T-1} - \hat{r}'_{T-1}u_{T-1}$$

where $\hat{r}_{T-1} = E[r|W_{T-1}]$; this leads to the decision rule

$$u_{T-1} = \hat{R}^+_{T-1}\hat{r}_{T-1}(c_T - sW_{T-1})/\hat{\alpha}_{T-1}$$

where $\hat{R}^+_{T-1}$ is the pseudo-inverse of $\hat{R}_{T-1} = E[\hat{r}_{T-1}\hat{r}'_{T-1}]$ and $\hat{\alpha}_{T-1} = \hat{r}'_{T-1}\hat{R}^+_{T-1}\hat{r}_{T-1}$. Clearly this rule is not unique, since $\hat{R}_{T-1}$ is usually singular but it allows sequential updating of the conditional estimate $\hat{r}_{T-1}$. Another rule which builds caution into the decision is to allow more flexibility in the model by dropping the constraint $E[W_T] \leq c_T$ which implies a preference by the investor of short term decision making. In this case we minimize

$$E[W^2_T] = \text{var}\, W_T + (E[W_T])^2$$

to obtain the rule

$$u_{T-1} = -(V + mm')^{-1}msW_{T-1}.$$

In this case also a sequential updating would transform this rule as

$$u_{T-1} = -\hat{R}_{T-1}\hat{r}_{T-1}sW_{T-1}.$$

for which the conditional objective function is

$$E[W_T|W_{T-1}] = s^2W^2_{T-1}(1-\hat{\alpha}_{T-1})^2.$$

Some comments on these decision rules and their associated frontiers are in order. First of all, they suggest that only a small amount of foresight or adaptivity is needed in most cases considered here for the investor to choose a myopic policy which is optimal in the sense defined. Secondly, the quadratic frontier equations such as (4.3.25) may be estimated from empirical data on mutual fund returns provided we can efficiently estimate the parameters m, V and c_T. In order to specify the dynamic estimating equation we consider now a general mean variance model with time- varying parameters e.g.

$$\min \text{var } W_T = E[W^2_T] - (E[W_T])^2$$

such that

$$E[W_T] = c_T,$$
$$W_{T+1} = s_t W_t + r' u_t$$
$$e' u_t = A \text{ (constant) for all } t, \; 0 \le t \le T\text{-}1.$$

Here e' is a row vector with each element unity, so that $e'u_t = A$ for all t imposes a budget constraint on the amount invested in risky assets in each period. The minimal variance $\sigma^2_T{}^*$ then satisfies the following dynamic efficiency frontier

$$\sigma^2_T{}^* = (s_{T-1}W_{T-1})^2\alpha_{T-1} - 2\lambda_T s_{T-1}W_{T-1}\alpha_{T-1} - 2k_T s_{T-1}W_{T-1}\beta_{T-1}$$
$$+ 2\beta_{T-1}\lambda_T k_T + \alpha_{T-1}\lambda^2_{T-1} + \gamma_{T-1}k^2_T, \qquad (4.3.26)$$

where

$$\lambda_T = (g_{T-1}\gamma_{T-1} - \beta_{T-1}h_{T-1})/(\alpha_{T-1}\gamma_{T-1} - \beta^2_{T-1}),$$
$$k_T = (\alpha_{T-1}h_{T-1} - \beta_{T-1}g_{T-1})/(\alpha_{T-1}\gamma_{T-1} - \beta^2_{T-1}),$$
$$\alpha_{T-1} = m'_{T-1}R^{-1}_{T-1}m_{T-1},$$
$$R_{T-1} = E[r_{T-1}r'_{T-1}],$$
$$\beta_{T-1} = e'R^{-1}_{T-1}m_{T-1},$$
$$\gamma_{T-1} = e'R^{-1}_{T-1}e,$$
$$g_{T-1} = c_T - s_{T-1}W_{T-1}(1-\alpha_{T-1}),$$
$$h_{T-1} = A + s_{T-1}\beta_{T-1}.$$

This is the dynamic analog of the quadratic efficiency frontier, the steady- state version of which takes the simple form:

$$\sigma^2_T{}^* = (\alpha\gamma - \beta^2)^{-1}(\alpha - 2\beta c + \gamma c^2),$$
$$u^* = (\alpha\gamma - \beta^2)^{-1}V^{-1}[(m\gamma - \beta)c + (\alpha e - m\beta)], \qquad (4.3.27)$$

where only risky assets are considered. Let c be replaced by its proxy, the average market return $\bar{R}_M$ like the S & P 500 index and I_t be the set of all realized information up to the latest period t and $\hat{\theta}_t = E[\theta \mid I_t]$ be the updated value of the parameter estimate for θ denoting any of the parameters above e.g., α, γ, β, m, V and c. Then the steady-state frontier (4.3.27) would appear as follows:

$$\sigma^2_T{}^* = \left(\hat{\alpha}\hat{\gamma} - \hat{\beta}^2\right)^{-1}\left(\hat{a} - 2\hat{\beta}\hat{c} + \hat{\gamma}\hat{c}^2\right),$$

Likewise the dynamic frontier may incorporate information-adaptive adjustment. It is clear that this type of adaptivity could be interpreted in terms of Kalman filters by using various instrument variables underlying the information set I_t, which may include various extraneous and market- related information available to the investor. Furthermore if the steady state frontier is desirable on long run considerations information adaptivity may allow closeness to the steady state. A third aspect of the decision rules above is that the dynamic efficiency frontier could in principle be derived for each of the four system dynamics specified before, and since the various models are embedded, only the goodness of fit tests would determine which frontier equations fit the empirical data best.

Robustness issues.

One could advance two basic reasons for analyzing the robustness of the efficiency frontier. One provides a justification of using the steady state frontier equations (4.3.27) as an asymptotic approximation i.e., even if the parameters are varying over time one could define an asymptotic or stationary efficiency frontier characterized by a myopic policy. Bertsekas uses such an argument in defending stationary portfolios. Even with such an argument the problem remains how to efficiently estimate the multivariate parameters (m, V) from the time series data on returns. It is not very clear that one should require the unbiasedness criterion for the sample estimates of vector m for example and if one admits biased estimators, one need not be restricted to the class of linear estimators. A second reason that has been advanced recently is the empirical evidence of nonlinearities in the distribution of stock returns which may be due to a nonlinear and non-Gaussian stochastic generating mechanism.

Consider the stochastic process generating mechanism underlying the last of the dynamic equations in (4.2.21a) and for analytical convenience a continuous-time version is more useful e.g.,

$$\dot{W}_t = -(adt + dB_t)W_t,$$

where

$$a = 1 - s - m'u_t, \qquad dB_t = -d\varepsilon'_t u_t \tag{4.3.28a}$$

and u_t is expressed as proportions of W_t and it is assumed that B_t represents a stationary Brownian motion with mean zero and variance $(\sigma^2 t)$. It is true that the assumption on the stochastic process B_t here is quite strong, but even with such strong assumptions the problems of instability of the dynamic frontier cannot be ignored. Hence we consider a simple illustration. Let $u_t = u^\circ$ be a constant vector not varying in time, then a is a constant parameter and the stochastic differential equation (4.3.28a) has the solution

$$W_t = W_0 \exp[-(a + \sigma^2/2)t - B_t], \tag{4.3.28b}$$

since the stochastic process B_t grows like $(t \log\log(t))^{1/2}$ with probability one, the stability property of the solution is determined by the deterministic term in the exponent of (4.3.8b) i.e., the region of sample stability (which is different from the true model stability) is specified by

$$\hat{a} + \hat{\sigma}^2/2 > 0 \tag{4.3.28c}$$

where $\hat{a}$, $\hat{\sigma}^2$ are the sample estimates of a and σ^2. The variance of the W_t process can be explicitly computed as

$$\mathrm{Var}(W_t) = W_0^2\{\exp[-(2\hat{a} + \hat{\sigma}^2)t] - \exp(-2\hat{a}t)\},$$

where W_0 is the value taken by W_t at $t=0$ with probability one. Clearly we need the condition

$$2\hat{a} = 2(1 - \hat{s} - \hat{m}'u^0) > \hat{\sigma}^2 \tag{4.3.28d}$$

for the variance $\mathrm{Var}(W_t)$ to be stable as $t \to \infty$. Since we have

$$E[|W_t|^k] = |W_0|^k \exp[\tfrac{1}{2}kt(k-1)\sigma^2 - 2a],$$

therefore for the stability of $E|W_t|^3$, $E|W_t|^4$ we need the respective conditions

$$\hat{a} > \hat{\sigma}^2 \text{ and } 2\hat{a}/3 > \hat{\sigma}^2$$

It is apparent that as k increases, the stability region decreases and hence the satisfaction of the condition (4.3.28c) for stability of the mean does not guarantee that the variance and other higher moments would be stable. Since the estimates $(\hat{s}, \hat{m}, \hat{\sigma}^2)$ can be readily obtained from the empirical data on returns, one could statistically test if the condition (4.3.28d) holds for stability of the variance and if it does then the steady-state approximation of the mean variance frontier by the quadratic equation makes sense, otherwise not. In the slightly more general case

$$\frac{dW_t}{dt} + (a + h_t)W_t = 0,$$

$$W_t = W_0 \text{ at } t = 0,$$

$$h_t = -\varepsilon_t' u^0,$$

where the h_t process is assumed to be stationary, ergotic and Gaussian with a zero mean, the stability region for the k-th moment is given by the inequality $2a > kS(0)$

for k = 1,2,3,4 where $S(w)$ is the spectral density of the h_t-process. By using the estimates of the spectral density one could therefore test for the stability of the mean and variance functions of W_t. These stability results have two important implications for the recent empirical studies on the nonlinear behavior of rates of return. One is in relation to the hypothesis of Hsu *et al.* (1974) who suggested a normal process for the returns but with random jumps in the variance parameter occurring at discrete points in time. In terms of the above conditions (4.3.28) this implies that the stability region would tend to decrease further and in such situations the case of stationary or myopic portfolio policy becomes more weakened. There is a second point that there may be significant estimation risks associated with the statistical estimates of the mean variance parameters obtained from sample observations. In a Bayesian framework Bawa, Brown and Klein (1979) have shown that if the predictive distribution of portfolio returns is not normal, then the variance may prove to be a very incomplete measure of risk. As a matter of fact some empirical tests by Beedles (1979) show that the skewness and kurtosis measures are significantly different from zero implying asymmetry in the statistical distribution of returns. All these points suggest the need for selecting robust portfolio policies, which are less sensitive to nonlinearity and nonnormality of returns. One may suggest two important ways through which robustness could be built into the portfolio policy. One is through the estimation risk underlying the mean variance parameters (m, V) of the steady state frontier. A second approach is to adopt a minimax decision rule to define a minimax portfolio which minimizes in some sense the maximum risk in portfolio selection defined suitably.

The risk associated with the statistical estimation of multivariate parameters (m, V) may be analyzed in several ways of which the following will be discussed here e.g.,

(A) How to obtain an efficient estimate of (m, V) when the sample return vectors r_t are known to be drawn from a multivariate normal population? This question is of some importance since the sample mean vector although unbiased for m is not generally an optimal estimate from a decision theoretic viewpoint in the sense that it is inadmissible.

(B) How to specify the portfolio efficiency frontier, when viewing portfolio return as a linear regression on market return? This question is important because it leads to a time series approach to the computation of efficient portfolios from historical data.

(C) How to apply the random parameter model of econometric theory to the specification and estimation of the portfolio efficiency frontier? This leads to an approach whereby sequential updating and revision of parameter estimates may be incorporated.

(D) How to apply a minimax class of estimate for a generalized portfolio model, where the portfolio or (security) return is linearly related to a number of factors and not one?

Note that each of the questions above relates to the robustness issue of portfolio policy and it has dynamic implications, though we restrict here to the steady state efficiency frontier.

Efficiency of alternative Estimators
The portfolio model

$$\sigma^2{*} = \min \left\{ u'Vu \mid mu = c,\ e'u = \sum_{i=1}^{n} u_i = 1 \right\}, \tag{4.3.29a}$$

which generates the mean variance efficiency frontier by varying c, usually represents c by the mean return $E[R_M(t)]$ of the market portfolio. This suggests that the balance condition $m'u = c$ may be estimated in several alternative ways as follows:

(a) $\bar{x}'u = \bar{y}, \quad \bar{y} = \bar{R}_M(t)$ (4.3.29b)

$\bar{x}, \ \bar{y}$ are sample means from observations;

(b) $\hat{x}'u = \bar{y}, \quad \bar{y} = \bar{R}_M(t)$ (4.3.29c)

where $\hat{x}$ is the James and Stein estimator of m

(c) $y_{(T,1)} = X_{(T,n)}\beta_{(n,1)} + \varepsilon_{(T,1)}; \quad u = \beta, \quad X = (r_{ti})$ (4.3.29d)

(d) $y^* = \beta' x^*, \quad x = x^* + \eta, \quad y = y^* + \varepsilon = \beta' x^* + \varepsilon,$ (4.3.29e)

this is the unobserved variable model, where $\beta = u$ as in (4.3.29d) and x^* is the mean of the return vectors. The assumptions about the zero mean error components η, ε are $\text{cov}(x,\eta) = \text{cov}(x,\varepsilon) = \text{cov}(y,\eta) = \text{cov}(y,\varepsilon) = 0$ with η and ε mutually and serially independent. and

(e) $E[z \mid y) = \mu_z + V_{zy}V_{yy}^{-1}(y-\mu_y), \quad z = x'u; \quad y = R_M(t)$ (4.3.29f)

the density $p(z,y)$ is assumed normal with means μ_z μ_y and variances and covariances as V_{yy} V_{zy}, and V_{zz}. The use of the sample mean vector

$$\hat{x} = T^{-1} \sum_{t=1}^{T} r_t$$

as an estimator of m has been criticized on the grounds that it is inadmissible for particular loss functions when $n \geq 3$. Thus let d be an estimator of m with quadratic loss $L = (d - m)'(d - m)$ having the risk function $R = R(m,d) = E[L]$, then without assuming normality, one can show, under fairly general conditions that there exists an alternative estimator d such that for $n \geq 3$ it beats the sample mean estimator in the mean square sense for all m i.e.,

$$E[(d - m)'(d - m)] < E[(\bar{x} - m)'(\bar{x} - m)]$$

In other words the sample mean estimator $\bar{x}$ is inadmissible. By using the more generalized loss function

$$L = L(m,d) = (d - m)'V^{-1}(d - m), \tag{4.3.29g}$$

James and Stein (1961) showed that even for normal samples the nonlinear estimator

$$d = \hat{x} = \left[1 - \frac{k}{\bar{x}'S^{-1}\bar{x}}\right]\bar{x}, \tag{4.3.30a}$$

S is an unbiased estimate of V, $k = (n - 2/(T - n + 2)$ has smaller risk than the linear sample mean estimator $\bar{x}$ for all parameter vectors m. Two cautions are needed however before we apply the nonlinear estimator $d = \hat{x}$. First of all, note that we have for normal samples

$$E\left[\sum_{i=1}^{n}(d_i - m_i)^2\right] < E\left[\sum_{i=1}^{n}(\bar{x}_i - m_i)^2\right]$$

This does not imply that $E[(d_i - m_i)^2] < E[(\bar{x}_i - m_i)^2]$ for each $i = 1,2,\ldots, n$. As a matter of fact some of the individual mean square errors of the $d_i = \hat{x}_i$ will be greater than those of the $\bar{x}_i$ so that for some constant vector a, the linear combination $d'a$ will have a larger mean square error than $\bar{x}'a$. When this happens the portfolio model (4.3.29a) has to be reformulated suitably. For instance consider the extended model

$$\min_{u,d} u'Vu + E[L(m,d)] \tag{4.3.30d}$$

such that $u'd = \bar{y}$, $e'u = 1$, where the loss function $L(m,d)$ is of the form (4.3.29g) and solve it as a two-stage decision process. First, we obtain the optimum JS estimator $d = \hat{x}$ given in (4.3.30a) by minimizing the expected loss function and then in the second stage we solve for an optimal vector $u^* = u^*(\hat{x})$ by minimizing the portfolio risk $u'Vu$ for given $\hat{x}$. However if the first-stage estimator $d = \hat{x}$ is such that there exists a vector a of constants with the property

$$E[\{(d - m)'a\}^2] > E[\{(\bar{x} - m)'a\}^2], \tag{4.3.30c}$$

then we cannot select $u^* = a$ in the second stage. Let A be the set of all such constant vectors a for which (4.3.30c) holds. We reformulate the second stage decision problem as

$$\min_u u'Vu$$

such that $u'd = \bar{y}$, $e'u = 1$, $u \notin A$. This can be formulated as a nonlinear program which can be solved to obtain a new optimal solution $u^{**} = u^{**}(\hat{x})$, provided it exists. The distance between the two optimal vectors u^* and u^{**} may thus be used to measure the sensitivity of the JS estimator in the context of the portfolio model. A second note of caution before applying the JS estimator in the context of the portfolio model is to point out that this estimator is biased, since we have

$$E[\hat{x}] = m - E\left[\frac{1}{n-2+2K}\right], \tag{4.3.30d}$$

where K is a random variable which has a Poisson distribution with mean $\frac{1}{2}m'V^{-1}m$. The risk function for the estimator $d = \hat{x}$ is:

$$R(m,d) = n - \left[\frac{T-n+1}{T-n+2}\right](n-2)^2 E\left[\frac{1}{n-2+2K}\right].$$

This implies that for values of m close to zero the risk advantages of the JS estimator $\hat{x}$ over that of the sample mean estimator $\bar{x}$ may be considerable. The sample mean $\bar{x}$ is the maximum likelihood (ML) estimator for the multivariate normal distribution, and in the class of linear unbiased estimators of m, it is the most efficient. But since the JS estimator $\hat{x}$ belongs to the class of biased and nonlinear estimators, it can beat $\bar{x}$ in terms of efficiency by allowing some trade-off between bias and efficiency. Judge and Bock (1978) have discussed several simulation results.

The JS estimator can be applied in portfolio theory in a different way when we consider the time-series version (4.3.29d) first proposed by Stohr (1977), who showed that by minimizing the expected sum $E[\varepsilon'\varepsilon]$ of squared errors subject to $u'e = 1$, we obtain the same optimal allocation vector u^* as in the traditional portfolio model (4.3.29a) with c representing the average market return. This time series version leads to the constrained least squares (LS) solution, if the error vector ε has zero mean and variance $\sigma^2 I$, where σ^2 is a scalar and I the identity matrix. The unconstrained LS estimate of β is b:

$$b = (X'X)X'y, \quad \hat{\varepsilon} = [I - X(X'X)X']y.$$

The positive part JS estimator d is related to the LS estimator b by

$$d = \left[1 - \frac{k\hat{\varepsilon}'\hat{\varepsilon}}{y'y - \hat{\varepsilon}'\hat{\varepsilon}}\right]^{+} b \tag{4.3.30e}$$

where the optimum choice of the constant k is $k = (n\text{-}2)/(T\text{-}n\text{+}1)$ and the superscript "+" indicates the positive part. In the constrained case we have to minimize the mean square error (MSE) of the JS estimator d subject to the constraint $\beta'e = 1$. On writing the constraint as a prior information $R\beta = e$ where R is an n by n matrix with each element unity and assuming for simplicity the orthonormal case (i.e., $X'X = \mathrm{I}$ after reparametrization) the constrained JS estimator d may be derived after Judge and Bock (1978) as:

$$d = \left[1 - \frac{k\hat{\varepsilon}'\hat{\varepsilon}}{(b-e)'(b-e)}\right]^{+} (b - e) + e \tag{4.3.30f}$$

which has risk smaller than the LS estimator b for all finite $\delta = R\beta - e$. Note that this estimator differs from the case (4.3.30d) in two respects. First of all, we are estimating here the allocation vector $u = \beta$ under the constraint $R\beta = e$ and after estimation we can write the balance equation as $m'd = c$, where values of m and c must be preassigned. Clearly in the general case both m and c have themselves to be estimated from the data set (y,X) but this leads to difficult problems of nonlinear estimation which are yet unresolved. Secondly, a two-stage version of the decision model can be formulated as before i.e., in the first stage we obtain a JS estimator d in (8.3.30f) from the time series data (y,X) and given that as fixed, we solve for the nonlinear estimator $\hat{x}$ of the mean return vector from (8.3.30a). The iterations continue until they tend to be stabilized or converge to some limiting value. Only by simulation experiments one can determine the efficiency of such a two-stage procedure.

There is yet another way by which the two main implications of the JS estimator for the mean variance frontier can be explored. The first implication is that the estimator allows for a trade-off of small bias to gain more efficiency in the sense of significant reduction in MSE whenever possible. Secondly, as nonlinear estimator of the mean parameter vector m, the JS estimator $\hat{x} = F(\bar{x})$ is a nonlinear function of the ML estimator $\bar{x}$, such that for small values of $\lambda = m'm$ the gain in efficiency is very significant, but for $\lambda \to \infty$ it has the same efficiency as the ML estimator $\bar{x}$ which is the sample mean.

A simple way to analyze the trade-off between bias and efficiency is to reformulate the efficiency frontier equation (4.3.27) under the condition that the ML estimates $(\bar{x}, \bar{V})$ of (m,V) change to $(h\bar{x}, h^2\bar{V})$ where for simplicity h is taken to be a scalar. The balance condition then gets transformed to $\bar{x}'u = c/h$ and the portfolio variance becomes $h^2u'\bar{V}u$. Hence the efficiency frontier reduces to

$$\hat{\sigma}^{2*} = (\alpha\gamma - \beta^2)^{-1}[\gamma c^2 - 2\beta ch + \alpha h^2].$$

Clearly the optimum value of h is given by $\frac{\partial\hat{\sigma}^{2*}}{\partial h} = 0$, i.e., $h_{opt} = h^* = \beta c/\alpha$, i.e., $0 < h^* < 1$, if $0 \leq c \leq \alpha/\beta$. The improvement region specified by $0 < h^* < 1$ says that by accepting alternative estimators $\hat{x} = h\bar{x}$, $\hat{V} = h^2\bar{V}$ the investor could reduce the minimal variance considerably if h^* is lower than one. For $h^* = 1$, no such reduction is possible. Clearly in the more general case one may seek instead of the scalar h a suitable matrix H for specifying the improvement region.

A more satisfactory way to analyze the trade-off between bias and efficiency is to adopt the logic of the decision rule (DR) approach and restrict the decision vector u in the portfolio model to zero-order, first order and higher order decision rules e.g.,

$$u = u(\theta) = u_0 + L_u(\theta) + Q_u(\theta),$$

where $\theta = (m, V^{-1})$, $L_u(\theta)$ and $Q_u(\theta)$ are linear and quadratic functions respectively of θ. For zero order decision rule $u = u_0$ i.e., it is independent of u. For linear DR if θ is replaced by its ML estimates we get back the extended portfolio model given in (4.3.30b). But for quadratic and higher order decision rules we need to specify a more generalized nonlinear model. In particular one has to incorporate a JS estimator for the variance covariance matrix V also and this needs a separate loss function. It is clear that more applied work and simulation experiments are needed to test the sensitivity of decision rules of order higher than one.

Finally, we consider the conditional regression model (4.3.29f) where the bivariate density $p(z,y)$ is assumed normal. A Bayesian interpretation is very natural in this framework. The estimate of the conditional expectation E[z|h] in this model is given by

$$z^F = \hat{\alpha} + \hat{\beta} y \tag{4.3.31}$$

where

$\hat{\alpha} = m_z - S_{zy}S_{zy}^{-1}m_y$, $\hat{\beta} = S_{zy}S_{zy}^{-1}$,

m_z, m_y are the sample means of z and y,

S_{zy}, S_{zy} are the sample moment statistics for V_{zy}, and V_{zy} respectively,

and this form provides the basis for the linear capital market line in the capital asset pricing model. However there is nothing in this specification which distinguishes it from an average regression line. Just as an "average" production function is different from an efficient production frontier, the efficient capital market line is distinct from an average line. This is so for two major reasons. One is the existence of incomplete information in the market e.g., Merton (1980) has shown that the market portfolio will not be mean variance efficient in such a situation. Secondly, due to heterogeneous expectations investors may respond to market signals with unequal transition probabilities. To model this situation one may specify a dynamic view of the model (4.3.31) as

$$z_{t+1} = z_t + \xi_t \tag{4.3.32}$$

$$y_d = z_t + \varepsilon_t,$$

where ξ_t, ε_t are zero-mean Gaussian white noise with $\mathrm{var}(\xi_t) = Q_t$, $\mathrm{var}(\varepsilon_t) = R_t$ and $E[\xi_t, \varepsilon_t] = 0$. Then the estimate z^F in (4.3.31) can be expressed as

$$\hat{z}_{t|t} = \hat{z}_{t|t-1} + G_t(y_t - \hat{z}_{t|t-1}), \qquad \text{for } t=1,2,3,....T$$

where G_t is the Kalman gain and $\hat{z}_{t|t-1}$ is the prediction of z_t using the data up to (t-1). Since $\hat{z}_{t|t-1}$ is a one-step prediction and $(y_t - \hat{z}_{t|t-1})$ is the error between observed y_t and the prediction, the estimation equation (4.3.33) is in fact a prediction-correction formula with the Kalman gain G_t as a weight factor. From control theory it is well known that if the dynamic system (4.3.32) is observable and the initial value z_o of z_t is known with certainty, then the Kalman gain G_t converges to a constant G, and therefore the forecasting equation for the optimal estimates $\hat{z}_{t|t}$ becomes simpler e.g.,

$$\hat{z}_{t|t} = \hat{z}_{t|t-1} + G(y_t - \hat{z}_{t|t-1}),$$

$$\hat{z}_{t|t-1} = \hat{z}_{t-1|t-1},$$

$$\hat{z}_{o|o} = E[z_o].$$

For the time-dependent case (4.3.33) the Kalman gain G_t can be computed recursively as follows:

$$P_{t|t} = (I - G_t)P_{t|t-1},$$

$$P_{t|t-1} = P_{t-1|t-1} + Q_{t-1},$$

$$P_{o|o} = \mathrm{var}[z_o],$$

$$G_t = P_{t-1|t-1}(P_{t|t-1} + R_t)^{-1}, \quad t = 1,2,...$$

One could apply these Kalman filter models to empirical data on returns of mutual funds, which are portfolios themselves. Thus, by taking three groups of funds such as growth funds, balanced funds and income funds one could obtain the optimal estimates of the vector $\hat{z}_{t|t}$. In particular one could test for the switching between mutual funds in a Markovian sense, since it can be written in a constrained form as

$$z_i(t+1) = \sum_{j=1}^{n} a_{ji}(t)\, z_i(t) + \text{error term}$$

with

$$\sum_{j=1}^{n} a_{ji}(t) = 1, \;\; a_{ji}(t) \geq 0 \;\; \text{for all } t.$$

If the diagonal elements are dominant in the sense that $a_{ii} > \sum_{j \neq 1}^{n} a_{ji}(t)$, then there is no switching. An empirical estimate by Dumas and Sengupta (1991) for 27 mutual funds (i.e., 13 growth funds, 9 balanced and 5 income funds) over the period 1964-1983 produced the following transition matrix assuming it to be time-independent:

	Growth	Income	Balanced
growth	0.332	0.637	0.031
	(1.81)	(1.66)	(0.08)
income	0.158	0.700	0.142
	(1.24)	(2.67)	(0.50)
balanced	0.093	0.542	0.365
	(0.76)	(2.15)	(.135)

where t-values are in parentheses. It is clear that growth and balanced funds evidence a high degree of switching compared to the income funds and this switching phenomena requires an explanation. It may be partly due to the changes in the relative riskiness of the funds, which are captured by the time-varying updating equations for the variance of z_t and the Kalman gain matrix G_t. Alternatively it may be due to the linear specification when the true model is nonlinear in form. More of such empirical applications are needed before one could reject the linear capital market line hypothesis of market efficiency.

Using Box-Cox transformations.

As an empirical application we consider now the estimation of two efficiency frontiers, one linear and the other nonlinear by using Box-Cox extended power family transformation over the data set of ten randomly selected mutual fund returns for two periods 1945-64 and 1965-83. The major motivation for Box-Cox transformations is that in some specifications the dependent variable may not be normally distributed but there may exist a transformation such that the transformed observations are normally distributed. For example, consider the nonlinear model

$$y_t = \exp(x'_t\beta) \exp(e_t) \tag{4.3.34}$$

$t = 1,2,\ldots,T$ where y_t is the tth observation on a dependent variable, x_t is a K-element column vector containing the tth observation on some explanatory variables, β is a vector of K parameters to be estimated and the errors e_t are normally independently distributed $N(0,\sigma^2)$ with mean zero and fixed variance σ^2. Clearly the dependent variable y_t is log normally distributed and heteroscedastic with variances $V(y_t)$:

$$V(y_t) = [\exp(x'_t\beta)]^2 [\exp(\sigma^2) - 1], \qquad t = 1,2,\ldots,T$$

However on taking logarithms of the equations (4.3.34) one obtains:

$$\ln y_t = x'_t\beta + e_t \tag{4.3.35}$$

where the transformed dependent variable $\ln y_t$ is normally distributed, homoscedastic and a linear function of β and hence the application of least squares (LS) to (4.3.34) yields a minimum variance unbiased estimator for β. Box and Cox consider a more general case of finding a transformation of the dependent variable y that simultaneously linearizes the model, corrects heteroscedasticity and normalizes the distribution of errors. Thus one type of transformation is of the following form:

$$y_t^{(\lambda)} = x'_t\beta + e_t$$

where

$$y_t^{(\lambda)} = y_t^{\lambda} - 1 \quad \text{if } \lambda \neq 0, \tag{4.3.36}$$

$$y_t^{(\lambda)} = \ln y^t, \quad \text{if } \lambda = 0,$$

and e_t is $N(0,\sigma^2)$, i.e. normally independently and identically distributed. Here it is assumed that there exists a transformation of the dependent variable, of the form given in (4.3.36), such that the transformed dependent variable has three properties, i.e. it is normally distributed, homoscedastic and has an expectation that is linear in β. Clearly if $\lambda = 1$ we get the familiar linear model: $y = X\beta + e$, and the value $\lambda = 0$ yields the nonlinear model (4.3.35). Thus if the hypothesis $\lambda = 1$ is rejected one could be assured that the nonlinear transformation is worthwhile to explore. In a more general case the dependent variables may also be transformed e.g.

$$y_t^{(\lambda)} = \beta_1 x_1^{(\lambda)} + \beta_2 x_2^{(\lambda)} + \ldots + \beta_K x_K^{(\lambda)} + e_t$$

Two general features of this class of transformations are particularly useful in applied econometric work. First of all, it provides an important method of generalizing functional forms; thus the two functional forms of the efficiency frontier (one linear and the other nonlinear can be generated by this transformation and only the best empirical fit with the observed data would determine the optimal value of λ, which can then be used to discriminate between the two specifications. Secondly, given the optimal value of λ, the conditional maximum likelihood (ML) estimates of the parameters β can be used along with the standard statistical tests. For example, in the case of (4.3.36) the probability density function $p(y)$ of the observations can be written as

$$p(y) = J\left(2\pi\sigma^2\right)^{-\frac{T}{2}} \exp\left[-\left(2\sigma^2\right)^{-1}\sum_{t=1}^{T}[y_t^{(\lambda)} - x_t'\beta]^2\right]$$

where

$$J = \prod_{t=1}^{T}\left|\frac{\partial y_t^{(\lambda)}}{\partial y_t}\right|$$

is the Jacobian of the transformation from y_t to e_t. For a given value of λ the conditional ML estimator of β is the LS estimator

$$b_\lambda = (X'X)^{-1}X'y^{(\lambda)},$$

and the maximized log-likelihood can be written as:

$$\ln L(\lambda) = -\frac{T}{2}(1+\ln 2\pi) - \frac{T}{2}\ln \hat{S}_E^2(\lambda) - \frac{T}{2}\sum_{t=1}^{T}\ln\hat{S}_E^2(\lambda) + (\lambda-1)\sum_{t=1}^{T} y_t$$

where $S_E^2(\lambda) = e_i'e_i/T$ is the residual sum of squares from a fit using the transformed responses since e_t is the vector of residuals from the regression of $y^{(\lambda)}$ on X. Box and Cox use the following iterative method for computing the optimal value of the unknown parameter λ. First, we compute $\ln L(\lambda)$ for a range of values of λ say between -2.5 to +2.5. By plotting $\ln L(\lambda)$ against λ one can locate the approximate value of λ that maximizes $\ln L$ (i.e. the ML estimate of λ). Let $\hat{\lambda}$ be the ML estimator of λ and suppose we wish to test the null hypothesis H_0: $\lambda = \lambda_0$ (e.g. $\lambda_0 = 1$ may be a null hypothesis i.e. no transformation is required). Then the likelihood ratio test statistic for the hypothesis

$$G_0^2 = -2\left[\ln L(\lambda) - \ln\left(\hat{\lambda}\right)\right] \sim \chi^2(1,\alpha)$$

is asymptotically distributed as a chi-squared variate, where $\chi^2(\nu,\alpha)$ is the $100(1-\alpha)$ percentile of a chi-squared distribution with degrees of freedom ν equal to the number of components in λ.

In our empirical applications we have utilized the extended power family transformation for both the dependent and the independent variables:

$$y^{(i)} = \begin{cases} \lambda_1^{-1}\left\{\left[\left(R_j - R_F\right) + \lambda_2\right]^{\lambda_1} - 1\right\}, & \lambda_1 \neq 0, \\ \ln\left(y + \lambda_2\right), & \lambda_1 = 0. \end{cases} \tag{4.3.37}$$

Here $y + \lambda_2 > 0$ and we choose the value λ_2=1.0 for all our empirical regressions. Likewise the independent variable is of the form

$$x^{(i)} = \begin{cases} \lambda_1^{-1}\left\{\left[\left(R_M - R_F\right) + \lambda_2\right]^{\lambda_1} - 1\right\}, & \lambda_1 \neq 0, \\ \ln\left(x + \lambda_2\right), & \lambda_1 = 0. \end{cases} \tag{4.3.38}$$

where $x + \lambda_2 > 0$ and $x = R_M\text{-}R_F$ and $y = R_j\text{-}R_F$. The two specifications of the efficiency frontier are as follows:

$$R_j = R_F = \alpha_j + \beta_j(R_M\text{-}R_F) + u_j \tag{4.3.39}$$

$$(R_j\text{-}R_F)^2 = a_j - b_j(R_M\text{-}R_F) + c_j(R_M\text{-}R_F)^2 + v_j \tag{4.3.40}$$

where R_j denotes the rate of return on mutual fund j, R_F is the risk-free rate of return, R_M the rate of return on market portfolio e.g., Standard and Poor Composite 500 Index and (u_j,v_j) are zero-mean error terms.

The extended power family specification (4.3.37) allows more flexibility in specification, where the likelihood ratio test statistic can be directly used. Hence we apply this model for the two periods: period I (1945- 64) and period II (1965-83) to the linear (4.3.39) and nonlinear (4.3.40) efficiency frontiers respectively by varying λ_1 over the following values 1.15, 1.50, 2.0, 2.5 and 3.0 with λ_2 fixed at 1.0 for retaining nonnegativity of the two variables y and x. Tables 1-2 present the estimates by using the extended Box-Cox transformation, where $\hat{S}_E^2$ * denote the minimum value of the squared standard error of regression when the optimal value of λ_1 is chosen.

Two points come out very clearly from these tables. First of all, the squared standard error $\hat{S}_E^2$ * of regression is uniformly lower for the nonlinear frontier than the linear frontier and this is more so for the growth funds than the balanced funds. Noting that the nonlinear frontier is only a regression specification of the frontier equation (4.3.27) in terms of second order statistics, its superior performance raises doubts on the linear specification. Secondly, the null hypothesis $\lambda^*_0 = 1$ that no nonlinear transformation is needed is strongly rejected by the Chi-square statistic G^2_0 at 1% level of significance. Clearly this suggests the need to explore the role of nonlinearities in the specification and estimation of the portfolio efficiency frontier and perhaps relate it to the third and fourth moments of the distribution of returns. Recent models of ARCH (AutoRegressive Conditional Heteroscedasticity) and Generalized ARCH may be used in this setup. These nonlinear models in conditional variances are discussed in a later chapter.

Table 1

Estimates of the linear frontier by the Box-Cox extended power family transformation for the optimal value of λ_1.

Mutual Fund		Period I			Period II		
No.	Group	Constant Intercept	R_M-R_F	S_E^{2*}	Constant Intercept	R_M-R_F	S_E^{2*}
1	B	-0.001	0.607**	0.030	0.011	0.884**	0.083
2	B	0.017	0.694**	0.044	0.020	0.755**	0.108
3	B	-0.002	0.521**	0.032	0.0006	0.706**	0.031
8	B	-0.034**	0.762**	0.033	-0.038**	0.736**	0.075
10	B	-0.005	0.716**	0.038	0.004	0.752**	0.046
5	G	-0.011	1.025**	0.096	0.023	1.058**	0.073
6	G	-0.073	1.419**	0.160	-0.058	1.294**	0.158
9	G	-0.025	1.063**	0.061	-0.010	1.032**	0.086
4	M	-0.039	1.071**	0.075	0.025	0.863**	0.074
7	M	-0.061**	1.228**	0.047	0.020	0.931**	0.070

Note: 1. Funds are grouped as B for balanced, G for growth, and M for mixed.
2. One and two asterisks denote significant t-values at 5% and 1% levels respectively.

Table 2

Estimates of the linear and nonlinear frontier by the Box-Cox extended power family transformation for the optimal value of λ_1.

Fund Group No.	Period I Constant Intercept	R_M-R_F	$(R_M$-$R_F)^2$	$S_E{}^{2*}$	Period II Constant Intercept	R_M-R_F	$(R_M$-$R_F)^2$	$S_E{}^{2*}$
1 B	0.00007	-0.031*	0.472**	0.006	0.015	0.007	0.407	0.027
2 B	-0.005	-0.022	0.837**	-0.011	0.017	0.053	0.316	0.036
3 B	-0.002	-0.012	0.366**	0.007	0.004	-0.022**	0.391**	0.006
8 B	0.004	-0.057**	0.553**	0.009	0.004	-0.046*	0.621**	0.014
10 B	-0.0009	-0.035	0.669**	0.009	-0.0002	-0.011	0.639**	0.010
5 G	-0.006	-0.119	1.715**	0.041	0.013*	0.026	0.789**	0.022
6 G	-0.036	-0.696**	4.871**	0.108	0.027	-0.158	1.424**	0.068
9 G	-0.00004	-0.121**	1.433**	0.017	0.012*	-0.056	0.869**	0.022
4 M	0.009	-0.135	1.251**	0.031	0.015**	0.019	0.332	0.021
7 M	0.004	-0.195**	1.692**	0.015	0.007	0.001	0.782**	0.021

Note: 1. Funds are grouped as in Table 1.
2. One and two asterisks denote significant t-values at 5% and 1% levels respectively.

4.4 Concluding Remarks

Stochastic control theory has several broad implications for economic models, when one compares it with deterministic control. The first is the information content of the dynamic model. Stochastic control assumes the information available to be less than complete, hence the need for evaluating the influence of uncertainty or inadequate information. By allowing the parameters to be random, optimal stochastic control derives rules which are more cautious than a deterministic model. Thus robustness is a natural concern for stochastic control theorist.

Secondly, the statistical estimation of a control model need not be once for all with a fixed time horizon. An adaptive updating scheme is usually built in through a scheme known as Kalman filtering. Various types of adaptivity are available in stochastic control literature and they are most valuable in economic applications. Moreover, the distribution of the random variables in the problem and the nature of the stochastic process underlying a control model has significant effects on the selection of an optimal control rule.

Finally, the economist's response to risk and uncertainty is to restate the objective function of the policy maker to allow for risk aversion and caution e.g. a linear combination of mean and variance of the target variables. The statistician response is to develop estimates of parameters which are robust and also adaptive to new information e.g. Kalman filtering. As we have illustrated, these two responses can sometimes be profitably combined by using both risk aversion in the objective function and also a sequential or recursive updating of parameter estimates.

References

Bawa, V., Brown, S. and Klein, R. (1979), *Estimation Risk and Optimal Portfolio Choice*, Amsterdam: North Holland.

Beedles, W. (1979), "On the Asymmetry of Market Returns." *Journal of Financial and Quantitative Analysis*, vol 14, pages 653-660.

Bertsekas, D. (1976), *Dynamic Programming and Stochastic Control*, New York: Academic Press.

Box, G., and Jenkins, G. (1970), *Time Series Analysis,Forecasting and Control*, San Francisco: Holden Day.

Chow, G. (1975), *Analysis and Control of Dynamic Systems*, New York: John Wiley.

Dunas, E.B., and Sengupta, J.K., (1991), "Nonparametric tests of portfolio efficiency under static and dynamic conditions." *International Journal of Systems Science*, vol. 22, pp 1929-1939.

Gasmi, F., Laffont, J. and Vuong, Q. (1990). "A Structural Approach to Empirical Analysis of Collusive Behavior." *European Economic Review*, vol.34, pages 513-523.

Holly, S. And Hughes-Hallet, A., (1989), *Optimal Control, Expectations and Uncertainty*, Cambridge University Press.

Hsu, D. and Wichern, D. (1974), "On the stable Paretian Behavior of Stock Market Prices." *Journal of the American Statistical Association*, vol. 69, pages 108-113.

James, W. and Stein, C. (1961), "Estimation with Quadratic Loss." *Proceedings of the 4th Berleley Symposium on Mathematical Statistics and Probability*, vol.1, pages 362-379.

Judge, G. and Bock, M. (1978), *The statistical Implications of Pre-test and Stein-rule Estimators in Econometrics*, Amsterdam: North Holland.

Kumar, P. And Varaiya, P., (1986), *Stochastic Systems, Estimation, Identification and Adaptive Control*, Englewood Cliffs, Prentice Hall.

Mangel, M., (1985), *Decision and Control in Uncertain Resource Systems*, New York: Academic Press.

Merton, R.C. (1980), "On Estimating the Expected Return on the Market: An Exploratory Investigation," Journal of Financial Economics, vol. 8, pp. 323-361.

Mossin, J. (1968), "Optimal Multiperiod Portfolio Policies." *Journal of Business*, vol. 41, pages 215-229.

Phillips, A.W., (1954), "Stabilization Policy in a Closed Economy." *Economic Journal*, Vol. 64, pp 290-323.

Roy, A., Hanssens, D. and Raju, J. (1994), "Competitive Pricing by a Price Leader." *Management Science*, vol. 40, pages 809-823.

Sengupta, J.K., (1985), *Information and Efficiency in Economic Decision*, Dortrecht: Martinus Nijhoff publishers.

Sengupta, J. and Sfeir, R., (1983), "Control Theory Models in the World Coffee." *International Journal of Systems Science*. vol.14, pages 811-827.

Stohr, E. (1977), "A Time Series Approach to the Computation of Efficient Portfolio from Historical Data." Discussion Paper, Center for Management Studies, Northwestern University.

Tintner, G. and Sengupta, J.K., (1972), *Stochastic Economics*, New York, Academic Press.

CHAPTER 5: **Economic Implications of Stochastic Control**

5.0 Introduction

The stochastic aspects of control theory raise some fundamental issues in modeling and estimating economic systems. Reasons are twofold. Economic data represent observed behavior, unlike the physical and engineering systems which represent regulatory and experimental mechanisms. Thus in servomechanism and feedback systems physical realizations due to control may be directly analyzed. But in economic systems this is almost impossible. Secondly, the objectives of performance are more implicit in economic systems. With several agents these objectives may in many cases differ to a significant degree; hence there is the need to analyze the problems of coordination in order to achieve some measure of consistency among different policies. We would analyze here some of these fundamental issues of control theory as applied to economic models.

5.1 Time inconsistency problems

Consider a dynamic model of an economy as a system of differences equations, where there are N decision agents, each controlling a vector $u_t(i)$ at time t with $i = 1,2,\ldots,N$.

$$x_{t+1} = f(x_t, y_t(1), \ldots, u_t(N), z_t) \,. \tag{5.1.1}$$

The state of the economy is here represented by the vector x_t which is not directly observable, so we have to use the output y_t which is observable as:

$$y_t = g(x_t) \,. \tag{5.1.2}$$

The sequence of variables $\{z_t\}$ over time denotes shocks generated by some stochastic process. It is clear that the system defined by (5.1.1) and (5.1.2) may well represent, under suitable qualifications, a typical econometric model of the Keynesian system. For example if both f and g are linear functions, x_t, y_t may, by suitable transformation include higher order lags and the shocks z_t are independent of u_t's with finite means and variances for all t, then this is a traditional econometric model, whose parameters θ can be estimated by least squares (LS) or other suitable techniques.

In an optimal control problem, distinct from the estimation problem, the decision agent i chooses the control vector $u_t(i)$ so as to minimize the expected value of a loss function, $E[L_i | I_{it}]$ conditional on his information set I_{it}, where

$$L_i = L_i(u_t(i), u_{t+1}(i), \ldots, x_{t+1}, x_{t+2}, \ldots) \tag{5.1.3}$$

and future control vectors $u_{t+k}(i)$, $k > 1$ are chosen in a similar fashion by minimizing the same criterion. Note that this optimal control problem is dependent on two basic premises: one is that the stochastic processes generating z_t and $u_t(j)$, $j \neq i$ are completely known as of now and for the future. For otherwise the optimal solution $u_t(i)$ would depend on the current information set I_{it} of agent i about the process $P(z,u)$ generating z_t, $u_t(j)$, i.e.,

$$u_t(i) = h_i^0(I_{it}),\ I_{it}: \{P(z,u),\ i \neq j\}\ . \tag{5.1.4}$$

It is clear therefore that if these generating processes P change any time in future and the agents are aware of it, their optimal control solutions $h_i^0(\cdot)$ would change. This is a basic problem of inconsistency, since in the real economic world models are only approximations, and even as approximations, future knowledge about the generating process P are far more uncertain. The second premise underlying the optimal control law (5.1.4) may be understood better if we write the optimal control law more explicitly as dependent on the set of parameters ϕ say:

$$u_t(i) = h_i^0(I_{it}, \phi) \tag{5.1.5}$$

where ϕ may be called the parameters of the optimal controller $u_t(i)$, which may be distinguished from the parameters θ of the dynamic system representing the econometric model specified in (5.1.1) and (5.1.2). For certain types of models known as LQG (Linear Quadratic Gaussian) control, there exist, under suitable assumptions on the generating process P, a mapping from the set θ to ϕ, where a separation principle holds. The separation principle implied by the optimal control rule (5.1.5) states that it is possible to make a separation between the estimation problem for θ in the econometric model (via maximum likelihood or Kalman-filtering methods of state estimation) and then the determination of the parameters in the optimal controller, which may be functions of the uncertainties (i.e., standard errors) of the econometric parameters, e.g.,

$$\phi = \phi(\hat{\theta}, F(\hat{\theta} | \theta)) \tag{5.1.6}$$

where $\hat{\theta}$ is an estimate based on the information set $I_t = \{I_{1t}, I_{2t}, \ldots, I_{Nt}\}$ and $F(\hat{\theta})$ is the probability distribution of the estimate $\hat{\theta}$ given the true value θ. Note that the role of information transfer is very critical here; the passage from $\hat{\theta} = (\hat{\theta} | I_t)$ to

ϕ in the optimal controller characterizes several forms of the separation principle. Three forms have been most widely used. One is the strong form of separation between estimation and control, also known as the *certainty equivalence* (CE) principle; the second is the weak form where the uncertainty in the estimate $\hat{\theta}$ is explicitly recognized. The third form uses separation as a principle of approximation of nonlinear models, more nonlinear than the LQG models and *suboptimal controllers* are used as first approximations which can be later improved with more information. The theory of *adaptive control*, where various forms of adaptivity and caution are built into, provides in a sense a general operational framework for dealing with cases where the separation principle, in weak or strong forms do not hold. Thus, in adaptive control which is the most active field of applied research in modern control theory there are very few cases where the CE principle is applicable. One exception is when the unknown parameters θ are stochastic variables which are statistically independent between different sampling intervals and other conditions of the LQG model apply. It is easy to show that in default of statistical independence, a simple random parameter model under LQG would fail to uphold the CE principle. Consider for example the following LQG model for one agent:

$$\min L = \mathrm{E}_0\left[\sum_{t=1}^{T} x_t' Q_t x_t\right]$$

subject to (5.1.7)

$$x_t = A_t x_{t-1} + B u_t + \varepsilon_t$$

$$x_0 \text{ unknown}, \qquad \varepsilon_t \sim \text{iid } N(0,\mathrm{I})$$

where $\mathrm{E}_0[\cdot]$ denotes expectation conditional on all information available up to the end of period zero, A_t, B are random parameters, with A_t time-varying and B time-invariant, Q_t is known but time-varying while the stochastic error term ε_t is assumed to be identically and independently distributed as $N(0,\mathrm{I})$. We follow the dynamic programming algorithm to solve the quadratic problem (5.1.7) for the last period, given the information up to the end of period T-1, i.e., we minimize

$$L_T = \mathrm{E}_{T-1}[x_T' Q_T x_T]$$

$$= \mathrm{E}_{T-1}[x_{T-1}(A_T' Q_T) x_{T-1} + u_T'(B' Q_T B) u_T + 2(u_T' B' Q_T A_T) x_{T-1}] + \mathrm{E}_{T-1}[\varepsilon_T' Q_T \varepsilon_T]$$

By setting to zero the first derivative of L_T with respect to the control vector we obtain the optimal decision rule

$$u_T^0 = G_T x_{T-1}; \quad G_T = -(\mathrm{E}_{T-1}[B' Q_T B])^{-1} (\mathrm{E}_{T-1}[B' Q_T A_T]) \,. \tag{5.1.8}$$

Following the recursive algorithm we may thus derive the optimal decision rule for any t > 0:

$$u_t^0 = G_t x_{t-1}; \quad G_t = -(E_{t-1}[B'Q_tB])^{-1} (E_{t-1}[B'Q_tA_t]) \tag{5.1.9}$$

where $E_{t-1}[\cdot]$ is expectation conditional on all information available up to the end of period (t-1). Let $\bar{A}_t$, $\bar{B}$ be the expected value of the corresponding random parameters and $\tilde{A}_t = A_t - \bar{A}_t$, and $\tilde{B} = B - \bar{b}$ be the deviations from their respective means. Then, on replacing the random variables by their respective means we obtain for the last period T, from the optimal rule (5.1.8):

$$G_T = \bar{G}_T = -(\bar{B}'Q_T\bar{B})^{-1} (\bar{B}'Q_T\bar{A}).$$

This is the optimal G_T for the CE principle, if it were correct to apply. However the true optimal rule requires a different G_T as:

$$G_T = -(\bar{B}'Q_T\bar{B} + E_{T-1}[\tilde{B}'Q_T\tilde{B}])^{-1} (\bar{B}'Q_T\bar{A} + E_{T-1}[\tilde{B}'Q_T\tilde{A}_T])$$

which indeed gives the correct optimal rule (5.1.8).

It is of some importance to analyze the various implications of these twin problems of inconsistency and separability for economic theory and policy modeling.

Consider the inconsistency problem first. We have in the optimal control model defined by (5.1.3) and (5.1.4) more than one decision agent, where their current decisions depend in part on their expectations of future policy actions. Thus if there are two agents, the private and public sectors for example, the decisions of the private sector would depend in part upon its expectations of future government policies. Unless these expectations are invariant to future policy changes, the optimal control rules calculated as of today would be inappropriate and inconsistent.

A. This type of inconsistency has been analyzed in some detail by Kydland and Prescott (1977), who have shown that in aggregate economic models, where there is usually a lack of details of the micro-structure due to inadequate data, the scope of such inconsistency may be compounded. The standard practice of using econometric models for choosing optimal government policy has followed a two-stage procedure: first one estimates an econometric model and then formally or informally uses optimal control theory to determine optimal policy, e.g., fiscal and monetary policy. But changes in policy induce change in the structure of the model which in turn necessitate reestimation and so on. This iterative process from structure estimation to control and then to reestimation and to control may not always converge. In this case the use of optimal control rules may even increase economic fluctuations rather than decrease them.

B. When the decision agents are several, the concept of a solution is no longer very precise, since various types of non cooperative solutions of differential games, e.g., Cournot-Nash, Stackelberg, etc. may be considered along with the cooperative solution concept. In the presence of uncertainty, the variety of possible solutions that one may envisage increases considerably. For instance, in a Cournot-Nash type of equilibrium solution, each player may have rational expectations in the sense that the expectation of the others' actions turns out to be the actual outcome, yet in default of appropriate coordination the sequence of revisions of a player's reaction curve may not always converge, the iterative process may then be as hazardous as optimal control rules that ignore possible changes in structure through expectations mechanisms.

C. The optimal plans in the form of open loop control policies for periods 1 through T require a commitment to the given horizon as viewed from the initial position. Thus if one plan at time τ does not turn out to be the same when formulated at $\tau + 1$, then the optimal policy may not be consistent. For instance, $\tau + 1$ may represent the date when a new tax on capital gains has been imposed with τ representing the no tax situation. Since policy interventions by the government in the form of new tax would change some of the constraints faced by the private sector, any policy model which calculates optimal policy without taking into account the reaction of the private sector and their constraint changes may be seriously flawed, depending on the magnitude of the intervention of course. Note that this type of inconsistency of optimal plans can arise even when the objective function (or the loss function) of the government is identical to that of the private sector in the form of a representative agent.

Next consider the problem of separability of estimation and control. There are two major consequences, when this separability fails to hold. One is that the conditional argument which is used either to forecast any *future* state given the *current* information, or to attain a *future* realizable target through current control, may not hold except perhaps very approximately and even in the latter case, one has to perform robustness and/or sensitivity analysis to test if the results are general. A second implication of the failure of separability is that the optimal estimation problem becomes jointly dependent on the optimal control problem, so that one has to unscramble so to say this interdependence before an optimal controller can be designed or applied. Sometimes this interdependency may be the consequence of lack of normality of the error structure in the dynamic plant equations.

It is useful to mention now some practical methods of resolving the time inconsistency problems. Note that there are many economic situations where the problems of inconsistency would not be very serious at all. For instance, in stable environments where growth of real income is associated with price stability, expectations of the private sector would not be much affected by the demand management policies of the government sector following a Keynesian strategy. Again in less developed countries where the assumptions of competitive market structure and price flexibility required by the neoclassical model may fail to hold, interventionism in government policy may have additional justification through the possibility of market failures and other bottlenecks.

In recent times several methods have been proposed both by economists and the control theorists to tackle the problems of inconsistency and inseparability. For the control through the physical systems and plant dynamics, these methods are known by their generic name of *adaptive control* involving self-tuning regulators which are discussed elsewhere. Of several methods of adaptive control, optimal dual controllers are assigned to utilize learning and caution along the way where future information is likely to be available for improving the control rule sequentially. However, these methods of adaptive dual control are not always very simple to apply due to computational difficulties, except in very simple cases. Also there are methods using differential game theory where two or more players are incorporated into the dynamic control model. Here the difficulty is that there may exist several types of non cooperative equilibrium solutions and hence there is no uniqueness.

5.2 Short-run vs. long-run optimality

For a deterministic control model with well defined objective functions, future expectations play no role. However most economic data involve future expectations about the state of the world and the dynamic model needs to specify how expectations are formed and how the agents adjust them over time. Precommitment to one fixed model without any adjustment leads to several types of biases. One type of bias is that it excludes the possibility of adaptive or rational expectations, by which agents learn and modify their behavior. This aspect has been already discussed in the previous section. A second type of bias arises due to the existence of evolutionary information, due to which economic behavior frequently involves time lags in adjusting stocks to their desired levels. For example, a firm which finds that its current factor stocks are inconsistent with the long run equilibrium implied by the current relative factor prices will generally spread the planned adjustment to long run equilibrium over a period of time. Meanwhile the desired stock level may change, so that in general all the expected future values of the target variables will be relevant to the current decision. Finally, a firm's dynamic decision process involving a highly complex profit maximization problem generally yields a highly nonlinear Euler equation defining the optimal trajectory, which is very often sensitive to the precise assumptions regarding the functional form of the chosen model. To avoid these complexities applied econometricians separate the optimization problem into two parts: the first considers a problem which gives rise to what may be regarded as the long run equilibrium solution for the firm. Next, the dynamic elements are introduced by postulating an optimal adjustment rule towards this equilibrium or desired value. Thus Callen, Hall and Henry (1990) have recently applied this specification to study the manufacturing sector in UK in respect to production, sales and inventories. Their first step assumes a simple profit function (π) of the firm which is to be maximized:

$$\max \pi = px - c(u) - A(\sigma_x^2, S, x) - hS \tag{5.2.1}$$

where p is the price of output, x is sales, u is production with its costs $c(u)$, S is the stock or inventory level with unit costs h. $A(\cdot)$ represents expected costs associated with a stock out, which are an increasing function of the level x and variance σ_x^2 of sales and a decreasing function of the level of stocks. The maximization in (5.2.1) is subject to the inventory constraint

$$S_t = S_{t-1} + u_t - x_t. \tag{5.2.2}$$

This step yields the optimal trajectory of stocks S_t^*, which is now taken to be the desired stock or inventory level. The second step then postulates a dynamic cost minimization problem, which imposes costs of deviating from S_t^* and also costs of adjustment in term of both the first and the second derivative as follows:

$$\underset{S_t}{Min}\, J = E\left[\sum_{t=1}^{\infty}[\,a(S_t - S_t^*)^2 + b(S_t - S_{t-1})^2 + c(\Delta S_t - \Delta S_{t-1})^2 | \Omega_t\,]\right] \tag{5.2.3}$$

where $\Delta S_t = S_t - S_{t-1}$ and Ω_t denotes the set of information available at t. Here the adjustment costs are assumed to be quadratic for reasons of mathematical convenience and no discounting factor is added since it is difficult to empirically estimate it. A similar model has been used by Kennan (1979), in a slightly generalized form, where a discounting factor is allowed for and the desired level of stocks is assumed to contain two parts: an expected part and a noise part.

We would now illustrate the above optimal decision rules in terms of the dynamic production and cost frontier model in the theory of the firm. These optimal rules specify the optimal profile of adjustment costs underlying a firm's intertemporal trajectory. A static production function assumes some inputs to be fixed in the short run (e.g., the capital inputs), and hence a static production frontier which is optimal in respect of the current inputs may not be optimal in a dynamic setting when the fixed or quasi-fixed inputs are also varied by the firm to achieve its long run goal. Furthermore, the dynamic production frontier must serve to characterize an optimal time path of investment in the fixed inputs. Once the firm's investment policy is optimally determined in a dynamic setting and the capital inputs in the short run are always updated by optimal investment in each period, we have a production frontier which is both satistically and dynamically efficient. Thus the technical inefficiency in production may be due to an non optimal expansion of capital inputs in the dynamic case and an non optimal use of the current and the quasi-fixed inputs in the short run. Hence if we denote the production function by $f(v,x,\dot{x})$, where $v = v(t)$, $x = x(t)$ are the current and capital inputs and $\dot{x} \equiv \Delta x(t) = x(t) - x(t\text{-}1)$ is the increment of capital or investment, then the technical inefficiency is characterized by:

$$y(t) \le f(v(t), x(t)) \tag{5.2.4}$$

in the short run with $\dot{x}$ set equal to zero, and by

$$y(t) \le g(v,(t),x(t),\Delta x(t)) \tag{5.2.5}$$

in the long run where $\Delta x(t)$ is determined by an intertemporal long run goal of profit maximization or cost minimization.

The theory of adjustment cost is usually adopted to explain the divergence between the static and the dynamic production function. Thus Treadway (1969) and more recently Morrison and Berndt (1981) present the long run optimization problem facing the firm as one of choosing the vector $x(t)$ of quasi-fixed inputs and the vector $v(t)$ of variable inputs so a to minimize the present value of the cost of producing a given flow of output $y(t)$ subject to a production function constraint (5.2.5). The present value of cost at time $t = 0$ is given by

$$V(0) = \int_0^{\infty} \exp(-rt)\left[\sum_{i=1}^{m} w_i v_i + \sum_{i=1}^{s} q_i z_i\right] dt \tag{5.2.6}$$

where z_i $= \dot{x}_i + d_i x_i =$ gross addition to the stock of x_i
d_i = fixed depreciation rate on the i-th quasi-fixed input stock
r = firm's discount rate
w_i, z_i = input prices of v_i and z_i.

The production constraint (5.2.5) is specified by Treadway (1969) in a separable form

$$y(t) = f(v(t),x(t)) - C(\dot{x}) \tag{5.2.7}$$

where $y(t)$ is output net of adjustment cost $C(\dot{x}) \equiv C(\Delta x)$, where we specify that the quasi-fixed inputs are subject to increasing internal costs of adjustment, i.e., as purchases of each quasi-fixed input increase, the amount of foregone output rises implying that the fixed input is available at increasing unit cost. By contrast the variable inputs are available at constant unit prices. The adjustment cost $C_i(\dot{x}_i)$ satisfies the following conditions

$$C_i(0) = 0, \quad \partial C_i(\dot{x}_i) > 0, \quad \partial^2 C_i(\dot{x}_i) > 0\,. \tag{5.2.8}$$

On minimizing the present value $V(0)$ of cost in (5.2.6) subject to the production constraint (5.2.7) one could obtain the optimal time paths of the current and semi-fixed inputs, i.e., $[v^*(t),x^*(t),\Delta x^*(t);\ 0 < t < \infty]$, where the asterisk denotes optimal values. By substitution one obtains the dynamic production frontier as:

$$y^*(t) = f(v^*(t),x^*(t)) - C(\Delta x^*)\,. \tag{5.2.9}$$

Note that this production frontier (5.2.9) satisfies the dynamic efficiency criterion implicit in the Euler equation and the transversality conditions analyzed by Treadway (1969) and others. The steady state of this model yields

$$y^* = f(v^*, x^*) \tag{5.2.10}$$

where it is assumed that Δx^* is set equal to zero and the convergence $x^*(t) \rightarrow x^*$ is achieved. The detailed conditions of convergence of the path of semi-fixed inputs $x^*(t)$ have been analyzed by Treadway (1969) and others. If these conditions hold then the steady state version (5.2.10) of the production frontier can be interpreted as a static efficiency frontier. If the observed output y_j of firm j satisfies the inequality

$$y_j < y_j^*; \qquad j = 1,2,\ldots,n \tag{5.2.11}$$

we have a case of technical inefficiency. For this situation Farrell (1957) proposed the following linear programming (LP) test, assuming the production function to be linear (or loglinear): for each unit or firm k solve the LP model:

$$\min g_k = \sum_{i=0}^{m} \beta_i v_{ik} + \sum_{i=1}^{s} \alpha_i x_{ik}$$

subject to (5.2.12)

$$\sum_{i=0}^{m} \beta_i v_{ij} + \sum_{i=1}^{s} \alpha_i x_{ij} \geq y_j;\ j = 1,\ldots,n$$

$$\beta = (\beta_i) \geq 0, \qquad \alpha = (\alpha_i) \geq 0$$

for the parameters α^*, β^* and find out for the k-th unit, if the following efficiency condition

$$\beta^{*\prime} v_k + \alpha^{*\prime} x_k = y_k \tag{5.2.13}$$

holds, where $y_k^* = \beta^{*\prime} v_k + \alpha^{*\prime} x_k$ may be interpreted as the maximum potential output. If $y_k^* = y_k$, then the unit or firm k is efficient. Otherwise $y_k^* > y_k$ and it is inefficient.

Now we consider the optimization problem for a firm as a two-stage problem. The first stage determines an optimal investment policy for the semi-fixed inputs by minimizing the present value of adjustment cost, i.e.,

$$\min V(0) = \int_0^{\infty} \exp(-rt)\left[\sum_{i=1}^{s} C_i(\dot{x}_i)\right] dt$$

subject to

$$\dot{x}_i = z_i - d_i x_i \tag{5.2.14}$$

$$x_i(0) > 0$$

$$C_i(\dot{x}_i) = \delta_i \dot{x}_1^2 ; \qquad i = 1,2,\ldots,s .$$

Let $[x_i^*(t),\ z_i^*(t);\ 0 < t < \infty]$ be the optimal trajectories satisfying the above dynamic model (5.2.14). Then the second stage uses the optimal investments $z_i^*(t)$ to update $x_i^*(t)$ by the rule:

$$x_i^*(t) = (z_i^*(t) - \Delta x_i^*(t))/d_i$$

to specify a dynamic production frontier as:

$$y(t) = f(v(t), x^*(t)) - C(\dot{x}^*)$$

$$C(\dot{x}^*) = \sum_i C_i(\dot{x}_i^*) = \sum_i \delta_i \dot{x}_i^{*2} .$$

Let H be the Hamiltonian for the problem (5.2.14):

$$H = e^{-rt}[-\sum_{i=1}^{s} \delta_i (z_i - d_i x_i)^2 + \sum_i p_i (z_i - d_i x_i)] .$$

We then use Pontryagin's maximum principle to derive the necessary conditions for the optimal trajectories as follows:

$$\dot{z}_i = (r + d_i)z_i - d_i(r + d_i)x_i$$

$$\dot{p}_i = (r + d_i)p_i - 2d_i\delta_i(z_i - d_i x_i) \tag{5.2.16}$$

$$p_i = 2\delta_i(z_i - d_i x_i)$$

$$\lim_{t \to \infty} e^{-rt} p(t) = 0 .$$

Note that the optimal demand for investment given by the first equation shows phases of increasing investment $(\dot{z}_i > 0)$ when the capital stock is low but decreasing investment $(\dot{z}_i < 0)$ when the stock of x_i is very high. The third equation states that the shadow price of investment in semifixed input $\dot{x}_i$ equals the marginal cost of expansion at time t. However the second equation (5.2.16) shows that as $\dot{x}_i$ increases more and more, it tends to depress the optimal path of shadow price of investment. Clearly the first three equations of the system (5.2.16) can be combined to derive a linear differential equation system in two variables (z_i, p_i), which could be solved given the initial conditions and the transversality conditions. These optimal solutions would characterize the dynamic efficient investment frontier, since by the strict convexity of the adjustment cost function these conditions are also sufficient for the optimal control problem (5.2.14).

Some general comments about the cost minimization model above are in order. First of all, the adjustment cost functions $C_i(\dot{x}_i)$ may be considered in more general forms. In this case the optimal dynamic investment demand function of (5.2.16) would take the following form:

$$\dot{z}_i = [(r + d_i)\,\partial C_i(\dot{x}_i)/\partial \dot{x}_i] / \left[\frac{\partial^2 C_i(\dot{x}_i)}{\partial \dot{x}_i^2}\right].$$

For example if $C_i(\dot{x}_i) = \exp(\gamma_i \dot{x}_i)$, then $\dot{z}_i = (r + d_i)/\gamma_i$. Thus by assuming specific functional forms of the nonlinear adjustment cost functions, these dynamic optimal investment demand functions can be econometrically estimated against observed data. Secondly, the adjustment cost parameters δ_i tend in general to affect investment growth adversely. This follows from (5.2.16) since we have

$$\dot{z}_i = d_i(z_i - d_i x_i) + \dot{p}_i / 2\delta_i$$

provided we have $\dot{p}_i > 0$ on the phase space. Thirdly, the optimal control model is a simplified version of a capital stock adjustment model. Clearly different regions in the phase space can be identified, which can then be analyzed for convergence properties, by following the standard methods of optimal control theory. From an econometric standpoint the identification of these different regions in the phase space may pose a challenging task. Finally, we note that unless the necessary conditions (5.2.16) of intertemporal optimality hold, the production frontier defined by (5.2.15) is not dynamically efficient. In this framework it is clear that the standard Farrell efficiency may not satisfy the conditions of dynamic efficiency. This is due to the fact that the second equation of (5.2.16) which is also called the perfect foresight condition may require that the firms correctly anticipate the future time path of the shadow prices of investment.

5.3 Model of Adjustment Costs with Rational Expectations

Adjusting current factor stocks for the uncertain markets in future frequently involves time lags due to cautious attitudes. Models of inventories and production analyze this adjustment process in terms of a linear decision rule (LDR) approach which involves minimizing an intertemporal quadratic cost function. This function specifies the costs of deviation of current factor stocks from their desired levels. Thus a firm which finds that its current factor stocks are inconsistent with the long run equilibrium implied by the current relative factor prices will generally spread the planned adjustment to long run equilibrium over a period of time. Meanwhile the desired stock level of inputs may change, so that in general all the expected values of the target variables become relevant for the current optimal decision. Thus there is a close relationship between the expectation formation and the distributed lags in any dynamic models of firm behavior.

The LDR approach provides a simpler model of optimal distributed lags and the adjustment process than the theory of adjustment costs discussed in the earlier section. This is due firstly to the assumption that the costs of deviations from the desired levels can be approximated by quadratic functions and secondly to the updating rule which involves consistent forecasts of the future used to estimate the production and inventory behavior in manufacturing industries see, e.g., Hay and Holt (1975), Kennan (1979) and Sengupta (1995).

Consider first a simple example where we have one input $x(t)$ and one output $y(t)$, both in logarithmic units such that $x^*(t)$ is a stochastic target variable desired by the decision maker, since it yields the target output $y^*(t)$:

$$\begin{aligned} y^*(t) &= \alpha_0 + \alpha_1 x^*(t) \\ y(t) &= y^*(t) + e(t) \end{aligned} \tag{5.3.1}$$

where $e(t)$ is a disturbance term assumed to be white noise. We assume that the firm's objective is to minimize the discounted expectation of the stream of costs conditional on the current information set $\Omega(t)$, i.e.,

$$\min \mathrm{E}\left[\sum_{t=1}^{\infty} r^t \left\{a_1(x(t) - x^*(t))^2 + a_2(x(t) - x(t-1)^2 \middle| \Omega(t)\right\}\right]$$

where r is a known discount factor. The first component of this loss function is a disequilibrium cost, the second component an adjustment cost, where the ratio $a = a_1/a_2$ measures the relative importance of these costs. By invoking the certainty equivalence theorem due to Simon (1956) and Theil (1958), one could easily derive the optimality condition as

$$a\{x(t) - \mathrm{E}_t[x^*(t)] + (x(t) - x(t\text{-}1)\} = r\mathrm{E}_t[\Delta x(t\text{-}1)] \tag{5.3.3}$$

where $\Delta x(t) = x(t) - x(t\text{-}1)$, and $E_t[.]$ denote the conditional expectation based on the current information set Ω_t. Since the expectation variables in (5.3.3) are unobserved we introduce the rational expectations hypothesis so that expectations may be replaced by future realizations. Thus we obtain the optimal trajectory equation:

$$rx(t+1) - (1+a+r)x(t) + x(t-1) = -ax^*(t) = -(a/\alpha_1)(y(t) - \alpha_0 - e(t)) . \qquad (5.3.4)$$

By defining the long run target $d(t)$ as

$$d(t) = (1 - \lambda r)\sum_{k=0}^{\infty} \lambda^k r^k x^*(t+k) \qquad (5.3.5)$$

where λ is the stable real root $0 < \lambda < 1$ of the quadratic characteristic equation associated with the second order difference equation system (5.3.3), one could also rewrite the optimal trajectory equation as a linear partial adjustment rule as follows:

$$\Delta x(t) = (1 - \lambda)[d(t) - x(t-1)], \quad r \geq 1 . \qquad (5.3.6)$$

Several comments may be added about this adjustment process. First of all, the optimal input $x^*(t)$ may be solved from (5.3.4) and then the production function may be written as

$$y(t) = \alpha_0 + \alpha_1 x^*(t) + e(t)$$

where

$$x^*(t) = -(1/a)[rx(t+1) - (1+a+r)x(t) + x(t-1)] .$$

Here the convergence of $x(t)$ to $x^*(t)$ as $t \to \infty$ is guaranteed since the stable root only is selected in (5.3.6). Note however that the optimal trajectory equation (5.3.4) assumes the rational expectation hypothesis. If this hypothesis fails to hold, it would introduce additional errors to the production function above. Secondly, if the target levels $x^*(t)$ are stochastic, their consistent estimates have to be obtained from the observed data on output. In case of nonstationary time series, suitable error correction models involving cointegrated variables may be required. Kennan (1979) has suggested a general method of instrument variables estimation as:

$$x^*(t) = g(t) + hQ(t) + \varepsilon(t)$$

where $g(t)$ is a time trend and $Q(t)$ represents an observed exogenous variable, with $\varepsilon(t)$ being a disturbance term reflecting the influence of omitted variables on $x^*(t)$. Thirdly, the partial adjustment rule (5.3.6) may be interpreted in terms of an adaptive expectation hypothesis based on a geometrically distributed lag model. Clearly the above method can be easily generalized for the case of many inputs and hence we may say that a short run production frontier

$$y(t) = \alpha' \tilde{X}(t) + \varepsilon(t) \tag{5.3.7}$$

is consistent with the dynamic frontier if its input vector $\tilde{X}$ (t) satisfies the optimal trajectory conditions such as (5.3.4) and the disturbance term $\varepsilon(t)$ is independent of the values of $\tilde{X}(t)$. Note that this frontier may be implicit in the regression equation obtaining $\tilde{X}(t)$ as a function of the exogenous variables. Alternatively, one may apply the nonlinear maximum likelihood method to estimate the parameter vector α by assuming that the error process $\varepsilon(t)$ is normally distributed.

5.4 An Application to New Growth Theory: Dynamic Adjustments with Learning by Doing

The last decade has seen a very rapid rate of economic growth in the so-called newly industrializing countries (NICs) of the East Asian Pacific Rim such as Singapore, Korea (South), Taiwan and Japan. Recently two types of attempts have been made to account for this success story, which has been termed a "miracle" by Lucas (1993). One is the growth accounting approach in the Solow tradition, which has been recently applied by Pilat (1993) to explain the postwar growth process of Japan. Besides Japan's capability to adapt and assimilate modern technology, Pilat identified the fast growth of capital stock as the prime mover of Japan's rapid growth. However capital formation is the result of demand and supply side influences and the demand from Japanese companies in the post war period was more dominant than the supply and this dominance played a major role according to Pilat. The standard neoclassical models of growth fail to emphasize this fact. The second type of attempt to explain the episode of rapid economic growth adopts a disequilibrium approach due to Benassy (1986), Lucas (1993), and others, where learning by doing and scale economies are directly emphasized as disequilibrating forces. In these attempts different economies reach different steady states, because of differences in initial endowments or in the accumulation of physical and/or human capital. This key result differs from Solow's neoclassical model, where different economies converge in one steady state regardless of their initial endowments.

Our object here is to evaluate the returns to scale property of the aggregate production process underlying Japan's growth and to identify the sources of sustained growth due to the major inputs such as capital, labor and the externality factor due to exports. To this end a model of adjustment cost is employed, which explicitly allows for the dynamic expansion path planned by the production agent. This is attempted here by a dynamic control theory model for the producer as the dynamic agent. The model, which is based on the formulation of Kennan (1979) and more recently of Gregory, Pagan and Smith (1993) involves a two-step decision process. In the first step the producer as the dynamic agent decides the optimal inputs by minimizing a steady state cost function, which gives rise to the long run equilibrium or target value. The second step then postulates an optimal adjustment rule towards the equilibrium or target level.

The first step optimization problem is:

$$\min C_t = w_t' X_t \quad \text{subject to} \quad Y_t = F(X_t) \tag{5.4.1}$$

where X_t and w_t are the column vectors of m inputs and their prices and Y_t is the given level of output subject to the production function $F(\cdot)$, which may exhibit increasing returns to scale. Let X_t^* be the optimal solution and Y_t^* the associated output $Y_t^* = F(X_t^*)$, which specifies the allocative efficiency frontier.

The second step of the optimization model assumes a short run adjustment behavior for the producer, who finds that his current factor uses are inconsistent with the long run equilibrium path (X_t^*, Y_t^*) above as implied by the current relative factor prices and their expected changes in the future. Since the desired levels of inputs may change over time, the adjustment of producer's economic behavior frequently involves time lags in adjusting stocks to their desired levels. Thus all the expected future values of the target levels X_t^* of input demand, which implicitly depend on the expected levels of future prices become relevant to the current optimal decision. Thus, as Treadway (1969) and Kennan (1979) have pointed out, any model involving *optimal* distributed lags requires a theory of expectation formation. Recently Callen, Hall and Henry (1990) and Sengupta (1992) have introduced into the adjustment process a quadratic component of costs due to producer's risk averse attitude toward input and output fluctuations. Assuming a quadratic adjustment costs the second optimization step requires the producer to minimize the expected present value of a quadratic loss function as follows:

$$\min_{x(t)} E_t[L] \tag{5.4.2}$$

where $$L = \sum_{t=0}^{\infty} r^t [(x_t - x_t^*)' \Lambda (x_t - x_t^*) + (x_t - x_{t-1})' \Psi (x_t - x_{t-1})]$$

where $E_t[L]$ is expectation as of time t, $r = (1+\rho)^{-1}$ is the exogenous rate of discount, $x_t = \ln X_t$, $x_t^* = \ln X_t^*$ and Λ, Ψ are the matrices of non negative weights. The first component of the loss function (5.4.2) is due to deviations from the steady state equilibrium, i.e., disequilibrium cost and the second component characterizes the producer's aversion to fluctuations in input levels, i.e., the smoothness objective. The first is an inefficiency cost incurred due to deviations from the targeted level of optimal input usage. The second is purely an adjustment cost indicating the costs of successive movement toward the optimal input combinations. One can illustrate this optimal adjustment procedure by means of a single input case. For this purpose we write the dynamic control model as:

$$\min_{x_t} E_t \sum_0^{\infty} r^t \; [\delta\,(x_t - x_t^*)^2 + (x_t - x_{t-1})^2\,]\,. \tag{5.4.3}$$

The first order condition for this optimization problem yields the Euler equation:

$$\Delta x_t = r\mathrm{E}_t[\Delta x_{t+1}] - \delta(x_t - x_t^*) \tag{5.4.4}$$

where $\mathrm{E}_t[\cdot]$ denotes the expectation taken with respect to information available up to time t and the notation Δx_t indicates the first difference $(x_t - x_{t-1})$. We denote the stable root of the quadratic characteristic equation of (5.4.4):

$$rz^2 - (1+r+\delta)z + 1 = 0$$

by μ and then the solution of the Euler equation reduces to the following expression:

$$x_t = \mu x_{t-1} + (1-\mu)(1-r\mu)\mathrm{E}_t\left[\sum_{t=0}^{\infty} (r\mu)^t x_t^*\right]. \tag{5.4.5}$$

These two equations above form the basis of the econometric estimation techniques developed here. Two general methods of estimation are possible here. In the first, the expectation term $\mathrm{E}_t[\Delta x_{t+1}]$ is replaced by $\Delta x_{t+1} + \eta_{t+1}$, where $\mathrm{E}_t[\eta_{t+1}] = 0$ and then the equation (5.4.4) becomes

$$\Delta x_t = r\Delta x_{t+1} - \delta(x_t - \alpha z_t) + v_t \tag{5.4.6}$$

where the error term $v_t = r\eta_{t+1} + \delta e_t$ and $x_t^* = \alpha z_t + e_t$. Here the stochastic target variable x_t^* is assumed to be linearly related to an observed exogenous variable z_t and e_t is a white noise error component. We would assume here that the error components η_t and e_t are statistically independent and then it follows that $\mathrm{E}_{t-1}[v_t] = 0$. Hence one could apply the method of instrumental variables (IV) to estimate the coefficients of (5.4.6) consistently. Kennan has used time and several exogenous variables instead of only one, i.e., z_t.

A second method employs an assumption about how z_t is generated and then derives an estimating equation. For example assume that z_t follows a first order process:

$$(1-\mathrm{L})z_t = \varepsilon_t, \qquad \mathrm{E}_{t-1}[\varepsilon_t] = 0 \tag{5.4.7}$$

where L is the lag operator, i.e., $Lz_t = z_{t-1}$. This yields the estimating equation

$$\Delta x_t = (\mu-1)(x_t\text{-}1 - \alpha z_{t-1}) + (1-\mu)\alpha'\Delta z_t + (1-r\mu)(1-\mu)e_t \tag{5.4.8}$$

which can also be written in the form of a partial adjustment model:

$$x_t = \mu x_{t+1} + (1-\mu)\alpha z_t + (1-r\mu)(1-\mu)e_t.$$

In case z_t follows a second order process, i.e., it is integrated of order two: I(2)

$$(1-L)^2 z_t = \varepsilon_t, \quad E_{t-1}[\varepsilon_t] = 0 \tag{5.4.9}$$

then the estimating equation becomes

$$\Delta x_t = (\mu-1)(x_t\text{-}1 - \alpha z_t) + (1-r\mu)(1-\mu)\alpha\Delta z_t + (1-r\mu)(1-\mu)e_t. \tag{5.4.10}$$

Similarly if z_t follows a third order process, i.e., it is integrated of order 3 we could derive an estimating equation from

$$(1-L)^3 z_t = \varepsilon_t, \quad E_{t-1}[\varepsilon_t]= 0\,. \tag{5.4.11}$$

Kennan (1979) was essentially concerned with the first order case (5.4.7) but in our empirical application higher order lags are considered essential, since the condition $E_{t-1}[\varepsilon_t] = 0$ is not fulfilled by the first order model.

While these two methods of estimation can be employed to any given empirical data set, the simulation studies by Gregory, Pagan and Smith (1993) found the first method to be more robust. Hence we have used the estimating equation (5.4.5) in a modified form. The first modification is to define a transformed target variable d_t as

$$d_t = (1-r\mu)E_t\left[\sum_{t=0}^{\infty}(r\mu)^t x_t^*\right] \tag{5.4.12}$$

in which case the estimating equation (5.4.5) becomes

$$x_t = \mu x_{t-1} + (1-\mu)d_t\,. \tag{5.4.13}$$

The second modification is to replace x_t^* by an estimate $\hat{x}_t^*$ derived from the IV method of estimation by using several exogenous variables such as z_t. This estimator $\hat{x}_t^*$ is derived to be statistically consistent, since the higher order lags in z_t are considered till the consistency property is satisfied. Finally, the transformed target variable d_t in (5.4.12) is forward looking since the future values of x_t^* (i.e., future values of z_t) are allowed here. One could also consider past values of z_t and hence past levels of x_t^* to estimate a backward looking model of d_t. These two variants provide the forward looking and the backward looking methods of estimation.

Now consider the general case of several inputs and the quadratic loss function (5.4.2) specified before. For simplicity we assume that the production function is log linear in form, i.e., $y_t = f(x_t)$ where $y_t = \ln Y_t$ and $f(\cdot)$ is log linear. On applying the first order condition for the minimization problem in (5.4.2) one obtains the optimal linear decision rule (LDR) as follows:

$$[-(1/r)(P + (1+r)\,\mathrm{I}_m)Z + (1/r)Z^2 + \mathrm{I}_m]x_{t+1} = -\,(1/r)P\mathrm{E}_t x_t^* \tag{5.4.14}$$

where $P = H^{-1}G$, I_m is the identity matrix of order m, and Z is a backward lag operator. It is well known that the characteristic equation of this linear difference equation system (5.4.14) will have half of its roots stable and half unstable. Let μ be the square matrix of stable roots of the system. Then one could define a long run target input demand vector as d_t:

$$d_t = (\mathrm{I}_m - r\mu)\sum_{s=0}^{\infty} r^s \mu^s x_{t+s}^* . \tag{5.4.15}$$

Note that this is identical to the transformed target variable d_t derived in equation (5.4.12) for the scalar case. On using this in the optimal LDR equation (5.4.14) we derive the linear adjustment rule (LAR) of the transitional dynamics, which is equivalent to the specification (5.4.13) in the one input case:

$$\Delta x_t = \phi(d_t - x_{t-1}) \tag{5.4.16}$$

where $\phi = \mathrm{I}_j - \mu$, d_t is derived in (5.4.15) and it is assumed that the rational expectations hypothesis holds, i.e., $\mathrm{E}_t[x_{t+1}] = x_{t+1}$. This assumption implies that the future expectations $\mathrm{E}_t[\cdot]$ are realized in terms of the observed variables x_{t+1}. Under conditions of perfect foresight and the possibility of no arbitrage such an assumption appears reasonable and it has been frequently applied in recent intertemporal macroeconomic literature.

The equations for the disequilibrium transition dynamics (5.4.16) can be estimated in principle by two different ways, as we have shown before. One is the forward-looking approach, where the future values of $x^*_{t+1} = h\cdot W_t + \eta_t$ with W_t as the instrument variables and η_t as the white noise error process are utilized in the estimation of the d_t series. The second is the backward looking approach where the past values x_{t-s}^* are used to estimate the d_t process, before the disequilibrium dynamics (5.4.16) is estimated. As we saw in the scalar cases (5.4.1) through (5.4.5) the estimating equations are different for different generating processes for the exogenous variables W_t, e.g., the case when W_t is integrated of order two differs significantly from the case when W_t is integrated of order one.

Since our interest is to evaluate the contribution of human capital, we consider heterogeneous labor inputs divided into three groups: F_t as total hours worked by female labor, U_t as total man-hours worked by male labor with lower skills (i.e., up to upper secondary level) and S_t as man-hours for skilled male labor with education levels higher than the upper secondary level. Also we have the capital input K_t representing physical capital in the input vector specification.

The modifications of the Solow model for explaining the episodes of very rapid economic growth in the Asian NICs usually take two forms. One is to redefine the aggregate production function in terms of output Y dependent on physical capital K, human capital H and augmented labor AL:

$$Y = F(K,H,AL)$$

where L is labor and A is an efficiency index representing technological innovation. In Arrow's learning by doing model the index A is related to the cumulative experience from the past which is proxied by a measure of capital stock, i.e., $A = A(K)$. Assuming constant returns and a Cobb-Douglas form, this version of production function (1) has been empirically applied by Mankiw, Romer and Weil (1992) over a set of cross-country data for 2 OECD countries, 98 non-oil and 75 intermediate countries over the period 1960-85.

The second specification, widely adopted in the growth accounting approach which has been recently surveyed by Norsworthy and Jang (1992) uses a production function of the form

$$Y = B\,F(K,L) \tag{5.4.17}$$

where B is an index of Hicks-neutral technical progress function. In recent empirical studies of Solow's productivity residual by Hall (1990) and others the technical index B is related to the externality factor V due to openness in trade and even the overall market power held by the producers in the industrial sector.

One may thus start with the log linear production function with four inputs: physical capital K_t, female labor F_t and two types of male labor, unskilled U_t and skilled S_t. However we do not assume a static specification of the aggregate production function such as (5.4.17), since it fails to capture the dynamic characteristics of the disequilibrium process underlying a phase of rapid economic growth. Instead we consider a dynamic specification of the production function in the Solow form as:

$$\Delta \ln Y_t = \alpha_0(V_t) + \alpha_1\Delta \ln K_t + \alpha_2\Delta \ln F_t$$
$$+ \alpha_3\Delta \ln U_t + \alpha_4\Delta \ln S_t + \alpha_5\Delta \ln V_t + \eta_t. \tag{5.4.18}$$

where η_t is the error process. This specification (5.4.18) is more flexible than (5.4.17) in several ways. For one thing, if the inputs and outputs are nonstationary but their first differences are stationary, i.e., integrated of order one, then this regression model has no strong bias towards accepting the spurious relationships. Also by the theory of cointegration one could formulate a suitable error correction model as follows:

$$\Delta y_t = \alpha_0 + \alpha' \Delta x_t - b_1(y_{t-1} - \beta' x_{t-1}) + \varepsilon_t \tag{5.4.19}$$

where $y_t = \ln Y_t$, $x_t = \ln X_{it}$ for the four inputs. If the disturbance term ε_t is white noise, then all the variables are stationary and hence standard regression estimates may be obtained with no difficulties of statistical inference. In the steady state we have the long run production relationship $y_t = \beta' x_t + \alpha_0/\theta$. A second flexible feature of the dynamic formulation (5.4.18) is that the intercept term $\alpha_0 = \alpha_0(V_t)$ may be replaced by a time-varying term $\Delta \ln R_t$ which indicates the adjusted Solow residual. The regression estimates of this adjusted Solow residual show if there is any procyclical variation over time. Finally, the dynamic specifications (5.4.18) and (5.4.19) imply that the production frontier can be alternatively expressed as optimal input demand functions derived before in (5.4.6) and (5.4.7). Whereas the production relation (5.4.18) provides a growth accounting approach, which decomposes output growth according to the growth of various inputs, the dynamic input demand models (5.4.6) and (5.4.7) specify the path of optimal changes in inputs along the expansion frontier. For the multiple input case these optimal input demand functions take the distributed lag form of equation (5.4.16) as we have shown before. This equation may be expressed as

$$\Delta \ln X_{it} = c_i - \phi_i \ln X_{i,t-1} + \beta_i d_{it} \tag{5.4.20}$$

where X_{it} denotes the four inputs: capital, female labor and two types of male labor, unskilled and skilled. Here the two coefficients ϕ_i and β_i are not ordinarily equal as the estimating equation (5.4.5) in the scalar case clearly shows. But here ϕ_i is equal to β_i due to two reasons. One is that the weighting matrices Λ and ψ in the objective function (5.4.2) are assumed to be diagonal. Secondly, the target variable d_t is defined here in a very special way. This is clearly shown in the derivations (5.4.12) and (5.4.13) of the scalar case. But there is a price to be paid for imposing this symmetry condition $\phi_i = \beta_i$ for each i through the transformed variable d_{it}. The method of estimation has to be a two-step procedure. In the first step we have to estimate the target input demand vector d_t from the intertemporal specification (5.4.11). But since this involves an infinite series which is convergent due to the choice of the stable roots, there are two ways of approximation. One is the forward looking model, the other is the backward looking model. The first maybe viewed as forward expectations, as viewed in Krugman (1991), whereas the second may be viewed as the role of past history. One could therefore test the existence of asymmetry in the roles of the future and of the past. The second step of estimation then applies the regression method to the linear equation (5.4.20) with $\phi_i = \beta I$ and d_{it} replaced by the estimate $\hat{d}_{it}$ from the first step.

For example in the backward looking specification we use the following linear estimating equation for capital

$$\phi_1 d_{1t} = \sum_{j=0}^{3}\left[\alpha_{1j}\ln\left(\frac{w_F}{w_K}\right)_{t-j} + \alpha_{2j}\ln\left(\frac{w_U}{w_K}\right)_{t-j} + \alpha_{3j}\ln\left(\frac{w_s}{w_K}\right)_{t-j} + \beta_j \ln Y_{t-j} + \gamma_j \ln V_{t-j}\right]$$

where w_K, w_F, w_U and w_s are the factor prices and Y_{t-j} and V_{t-j} are the real incomes and real exports lagged. Note that this approximation assumes that d_{it} can be explained by the lagged values of the exogenous variables represented by the relative factor prices, national income and exports. Similar approximations for the other three inputs (F,U,S) can be made. In the forward looking model d_{it} is explained in terms of three future values of the exogenous variables at $t+j$, $j = 1,2,3$. Since we are interested in evaluating the role of human capital, it is useful to illustrate the two-step estimation procedure for a two input model as in (5.4.17) that is frequently applied in the growth accounting approach. With L_t as total labor and K_t as capital and a Cobb-Douglas production function the desired levels of the two inputs L_t^* and K_t^* are determined by minimizing the static cost function now written as

$$\text{Min } w_t L_t + g_t K_t \qquad \text{such that} \qquad Q_t = B(A_t, V_t) K_t^a L_t^b$$

where Q_t is output and w_t, g_t are the respective prices of the two inputs: labor L_t and capital services K_t. The technical progress variable $B(\cdot)$ is assumed to be a function of exports V_t and an external shock variable A_t representing the international transmission of knowledge. Allowing for stochastic disturbances ε_L and ε_K the first order conditions of minimum cost are

$$\ln\ L_t^* = \ln(a/b)^{a(a+b)^{-1}} + (a+b)^{-1}[\ln B_t - a\ \ln(w_t/h_t) + \ln Q_t] + \varepsilon_L \tag{5.4.21}$$

$$\ln\ K_t^* = \ln(a/b)^{a(a+b)^{-1}} + (a+b)^{-1}[\ln B_t - b\ \ln(w_t/h_t) + \ln Q_t] + \varepsilon_K .$$

Given these target levels of optimal inputs the producer minimizes the following loss function

$$\min_{x_t} E_t \sum_{t=0}^{\infty} r^t[(X_t - X_t^*)\Lambda(X_t - X_t^*) + (X_t - X_{t-1})'\Psi(X_t - X_{t-1})] \tag{5.4.22}$$

where r is the discount rate, $X_t = (\ln L_t; \ln K_t)$ is the vector of actual input levels and $X_t^* = (\ln L_t^*; \ln K_t^*)$ is the target levels determined above in (5.4.21). Assuming diagonal weight matrices one obtains

$$\Lambda = \begin{pmatrix} \Lambda_L & 0 \\ 0 & \Lambda_K \end{pmatrix}, \Psi = \begin{pmatrix} \Psi_L & 0 \\ 0 & \Psi_K \end{pmatrix}.$$

The first order conditions for minimizing the intertemporal loss function (5.4.22) may then be written as:

$$\Lambda_L (\ln L_t - E_t(\ln L_t^*)) + \Psi_L (\ln L_t - \ln L_{t-1}) - r\Psi_L(\ln L_{t+1} - \ln L_t) = 0, \quad (5.4.23)$$

and

$$\Lambda_K (\ln K_t - E_t(\ln K_t^*)) + \Psi_K (\ln K_t - \ln K_{t-1}) - r\Psi_K(\ln K_{t+1} - \ln K_t) = 0. \quad (5.4.24)$$

Since these two equations are of the same form, it is sufficient to analyze the labor demand equation only. By rearranging (5.4.23) one gets

$$\theta_L E_t(\ln L_t^*) = (\theta_L + 2)\ln L_t - (1/r)\ln L_{t-1} - \ln L_{t+1} \quad (5.4.25)$$

where the ratio $\theta_L = \Lambda_L/\Psi_L$ measures the relative importance of the disequilibrium cost to the adjustment cost. On applying the lag operator Z to the right had side of equation (5.4.25) we obtain

$$-\theta_L E_t[\ln L_t^*] = \{1 - (\theta_L + 2)Z + Z^2\}\ln L_{t+1}.$$

Then by using the stable (i.e., non explosive) root μ_1 one could write this as:

$$(1 - \mu_1 Z)\ln L_{t-1} = (1 - \mu_1)d_t^L \quad (5.4.26)$$

where $d_t^L = (1 - \mu_1)\sum_{s=0}^{\infty} \mu_1^s \ln L_{t+s}^*$.

Define ϕ_L as $1-\mu_1$ and subtract $\ln L_{t-1}$ from equation (5.4.26). This yields finally the optimal adjustment rule for labor demand:

$$\Delta\ln L_t = \phi_L(d_t^L - \ln L_{t-1}). \quad (5.4.27)$$

Doing likewise for the capital input yields the adjustment rule for capital:

$$\Delta\ln Kt = \phi_K(d_t^K - \ln K_{t-1}). \quad (5.4.28)$$

Note also that one could derive here a relationship between the speed of adjustment parameter ϕ_i and the ratio θ_i of the weights on the two costs, i.e.,

$$\theta_i = -\phi_i r + \phi_i(1-\phi_i) - 1, \qquad i = L,K. \tag{5.4.29}$$

Note also that the equations (5.4.23) and (5.4.24) can be rearranged to define the following variables

$$Z_t^L = -\theta_L\left[\ln(a/b)^{-a(a+b)^{-1}} + h_B \ln B(A_t,V_t) + h_{(w/g)}^L \ln(w_t/g_t) + h_Q \ln Q_t\right]$$

$$Z_t^K = -\theta_K\left[\ln(a/b)^{-a(a+b)^{-1}} + h_B \ln B(A_t,V_t) + h_{(w/g)}^K \ln(w_t/g_t) + h_Q \ln Q_t\right] \tag{5.4.30}$$

where:

$$Z_t^L = \ln L_t - (\theta_L + r + 1)\ln L_{t-1} + \ln L_{t-2},$$
$$Z_t^K = \ln K_t - (\theta_K + r + 1)\ln K_{t-1} + \ln K_{t-2}.$$

The coefficients of (5.4.30) can be easily derived as follows:

$$h_B = (a+b)^{-1}, h_Q = (a+b)^{-1}, h_{(w/g)}^K = b(a+b)^{-1}$$

$$h_{(w/g)}^L = a(a+b)^{-1}. \tag{5.4.31}$$

Thus it is clear that if we can find estimates of the adjustment coefficients ϕ_L and ϕ_K in (5.4.27) and (5.4.28), then we can calculate θ_L and θ_K from (5.4.29). Then the coefficients h in (5.4.31) can be exactly identified and estimated by the method of ordinary least squares. Both Kennan (1979) and Gregory et al. (1993) have shown the optimum properties of the ordinary least squares method.

In our empirical application the term $B(A_t,V_t)$ is assumed to be of the form $A_t V_t^{\gamma}$, where A_t represents a time trend and V_t is the externality factor represented by the trend in real exports. We note that the method of estimation described above for the case of two inputs can be easily formulated for the four input case.

Some general comments may now be added here. First of all, we note that the estimate $\hat{\phi}_L$ of the adjustment parameter for the labor demand equation (5.4.27) is utilized here in three ways. One is to test the speed of convergence to the steady state level $\bar{d}_L$, which leads to the steady state level of output through the steady state production function. Secondly, the second step estimates of the adjusted demand functions provide an estimate of the scale economy of the aggregate production function. Thirdly, it provides an estimate of the ratio $\theta_L = \Lambda_L/\Psi_L$, which denotes the relative proportion of the two weights: Λ_L for deviations from the desired level of labor demand and Ψ_L for deviations from last year's level.

Moreover one has to note that the desired levels of input demand d_t^L and d_t^K in (5.4.27) and (5.4.28) have to be estimated by the IV method by using all the exogenous variables such as past incomes, relative factor prices and past exports in the backward looking approach. For the forward looking approach the same exogenous variables in their future values have to be utilized. By comparing these two approaches one could test the existence of asymmetry in growth response from the past history versus future expectations. In our empirical applications we applied AR(2) and AR(3) processes in the IV method of estimation but since the AR(3) process provided better values of DW statistics we retained only these in Table 4. These estimates are discussed in the next section along with their implications for the dynamic adjustment paths for the various inputs. The technical details of the estimation procedure and their asymptotic properties have been analyzed by Okamura (1994) separately.

For estimation of the production frontier models specified in (5.4.14) through (5.4.16) the aggregate data for Japan (1965-90) are utilized for labor, capital and their prices. The output Y_t is defined as GNP deflated by the GNP deflator and the external variable V_t is represented by exports deflated by the official export price index. These statistical data are obtained from the *Statistical Yearbook of Japan* for various years. It is difficult to obtain appropriate data for the capital stock and its price. The data for capital stock is used as a proxy for capital service and this is obtained from the official publication entitled *Gross Capital Stock of Private Enterprises* for various years. The data for labor services and wages are taken from the statistical report entitled: *Chingin-Kozo Kihontokei Chosa* for various years. Labor data is classified by gender and four educational levels: primary and secondary, upper secondary, junior college and four year college or above. With these data set the two-step growth model is estimated so that the estimates are statistically consistent. The estimation results reported in Tables 1-4 may now be discussed in three parts. First, we discuss the results of the production frontier estimates in a Solow-type framework, where the time-varying nature of Solow's productivity residual is analyzed. Second, the convergence of the time paths of inputs and output is discussed in terms of the speed of adjustment and other characteristics of the underlying dynamic system. Finally, the two-step estimates of the labor demand equation are analyzed in order to test the relative importance of the two costs of deviations: one from the desired level and other from the past levels.

The estimation results in Table 1 for the production frontier with heterogeneous labor have some interesting features. First of all, the role of capital is much more important than labor, at least twice or more in terms of the respective coefficients. This result is upheld even when a two input Cobb-Douglas production function is estimated. Using data from the UN Comparison Project (1960-85) De Long and Summers (1991) found a similar result on the dominant contribution of equipment investment to economic growth in Japan. They noted that the differences in equipment investment account for essentially all of the extraordinary growth performance of Japan relative to the sample of 25 countries over the period 1960-85. Secondly, the relative contributions of labor and exports are much smaller, implying that the phase of increasing returns is on the decline. Since exports is the only variable used as a proxy for eternality, it suggests a pattern different from the other NICs in Asia such as Korea and Taiwan for example. In 1986 the ratio of export to GDP was 17.2% in Japan, whereas it was 48.1% in Korea and 84.0% in Taiwan. This ratio fell to less than 12% in Japan in 1993. As a mature industrial country the impact of openness in trade on the overall economic growth in Japan is much less than the other NICs in Asia. Another factor may be the existence of strong diminishing returns in the learning by doing process as has been established by Young (1991) in his empirical studies. Thus in the initial stage the development of new technologies leads to rapid learning by doing. After some time however the productive capability of these new technologies is exhausted and learning by doing slows down. Finally, the process of the externality factor helps in improving the elasticity of output with respect to female labor, although the coefficient is not statistically significant at 5% level of t-test. This shows that the externality factor has been playing a declining role.

Note that the overall returns to scale (RTS) is increasing if the unskilled labor input which has a negative coefficient is excluded. Thus the estimated value of RTS is 1.023 when exports are not incorporated and it is 1.004 when it is included. But allowing for the negative coefficient of unskilled labor the RTS changes to 0.877 (without exports) and 0.857 (with exports) respectively. These results however provide a sharp contrast to the estimates of RTS obtained from a static form of the Cobb-Douglas production function, where the inputs and outputs are not in first difference terms. These latter estimates have consistently shown the persistence of strong increasing RTS for Japan.

The estimates of the adjusted Solow residual reported in Table 2 clearly show its procyclical variations. For the U.S. economy and its major industrial sectors Hall (1988,1990) found a similar pattern and attempted to explain it in terms of the market power and exogenous government expenditure. In our case the change in GNP is found to be positively correlated with the adjusted Solow residual. However one should note that the estimate of the residual is much less than that obtained by Norsworthy and Jang (1992). Two reasons may be given. One is that our estimates are derived directly from the aggregate production function and not from the shares of labor and capital in total national income. Second, the export variable and the heterogeneous labor have separate roles in the growth process and the elements of market power or the degree of monopoly may not be ruled out in this context.

The estimated results in Table 3 are of major significance, since the convergence to the steady state growth of the inputs and output is directly tested here in an econometrically satisfactory manner. First of all, when the externality factor is not introduced, the capital input has faster convergence to its steady state than the skilled female labor but this pattern is reversed when exports is introduced as an explanatory variable. Thus the presence of exports reduces the speed of convergence of capital relative to the major types of labor. Secondly, the unskilled manpower imparts oscillations in the convergence path, which may be indirectly responsible for the procyclical variations of Solow's productivity residual. Finally, one notes that the skilled manpower variable S (skilled labor) has the highest value of $\phi_1 = 0.826$ compared to that of capital and female labor, when the export variable is used in the specification. This suggests faster convergence to steady state of skilled manpower. For a mature industrial economy like Japan this may provide strong empirical support to the hypothesis that openness and trade help in the process of rapid development of skilled personnel to its desired target level. The so-called knowledge spillover hypothesis may be still active in the industrial process, although the learning by doing effects may be reaching stages of exhaustion. In this aspect of rapid deployment of skilled personnel in the growth process, Japan may be very similar to the other successful NICs in Asia. There exists some indirect empirical evidence analyzed by Sengupta (1993), which shows that the traditional export items were very dissimilar in pattern across the four countries: Korea, Hong Kong, Japan and Taiwan but similar and homogeneous patterns were found for exports of certain sophisticated R&D intensive products such as electrical machinery, telecommunications and computer equipment, etc. The results of direct regression estimates of the equation

$$\Delta GDP = a_0 + a_1 \Delta K + a_2 \Delta N + b \Delta V$$

reported in Table 5 show that the export externality factor ($\hat{b} = 0.12$ which is not statistically significant at 5% level of t-test) is much less important a factor compared to the other successful NICs in Asia such as Korea and Taiwan.

Finally, the two-step estimation results reported in Table 4 provide some interesting insight into the convergence of the optimal labor demand equation for the two-input model. Since the long run level d_t^L defined in (5.4.26) may be estimated either by a forward-looking or a backward-looking procedure one could derive two sets of estimates of the labor adjustment equation. The results of the backward looking model are reported in Table 4 and this is the conventional method followed by Kennan and others to derive statistically consistent estimates. However for a growing economy interested in sustained growth, the backward and forward looking models need not produce identical results. To test this difference we estimated the forward looking model for labor demand equations as follows:

First step

$$Z_t^L = -258.3 - \underset{(0.827)}{3.545} \ln Y_{t-1} - \underset{(2.708)}{5.496} \ln RW_{t-1}$$

$$R^2 = 0.608; DW = 1.557$$

Second step

$$\Delta \ln L_t = 16.6 - \underset{(0.18)}{0.811} \ln L_{t-1} + \underset{(0.50)}{1.117} \ln Y_t - \underset{(0.55)}{1.028} \ln Y_{t+2}$$
$$+ \underset{(0.21)}{0.447} \ln RW_t - \underset{(0.202)}{0.368} \ln RW(t+1)$$

$$R^2 = 0.635; DW = 2.18; \hat{\theta}_L = 3.489$$

First step

$$Z_t^L = -103.7 - \underset{(3.397)}{9.487} \ln Y_{t-1} - \underset{(1.694)}{1.527} \ln RW_{t-1} + \underset{(1.655)}{3.686} \ln V_{t-1}$$

$$R^2 = 0.655; DW = 1.749$$

Second step

$$\Delta \ln L_t = 15.7 - \underset{(0.123)}{0.837} \ln L_{t-1} + \underset{(0.476)}{1.081} \ln Y_t$$
$$- \underset{(0.714)}{2.460} \ln Y_{t+2} + \underset{(0.550)}{1.502} \ln Y_{t+3} + \underset{(0.151)}{0.457} \ln RW_t$$
$$- \underset{(0.147)}{0.783} \ln V_t + \underset{(0.175)}{0.484} \ln V_{t+1} + \underset{(0.135)}{0.405} \ln V_{t+3}$$

$$R^2 = 0.877; DW = 1.824; \hat{\theta}_L = 4.281$$

Here RW_t is the ratio of factor prices of labor to capital and d_t is the target levels of optimal labor demand.

Two points emerge very distinctly. One is that the forward looking estimates are very different in magnitude compared to the backward looking estimates. Furthermore future incomes have strong impact on the current demand for labor. When only the skilled component of the total labor demand equation is considered, this impact is reinforced more strongly. Secondly, the value of $\hat{\theta}_L$ as defined in (5.4.25) by the ratio Λ_L/Ψ_L is nearly half that of the backwarding model thus suggesting an asymmetry in the adjustment process. If the GNP variable Y_t is replaced by GDP the estimate of the ratio $\hat{\theta}_L$ is still very high:

Variables	Value of $\hat{\theta}_L$	
	Forward-looking	Backward-looking
ln *GNP* without exports	3.489*	13.944*
ln *GNP* with exports	4.281*	7.873*
ln *GDP* without exports	7.008*	13.687*

* Significant at 5% level.

For the two industrial sectors: durable and non durable manufacturing in U.S. over 1947-69 Kennan's two-step estimates consistently produced a value of $\hat{\theta}_L$ less than 0.14, i.e., 0.134 for the durable and 0.053 for the non durable sector. This provides a strong evidence that the U.S. producers place much more weight on smoothing out fluctuations in input demand, whereas their Japanese counterparts emphasize their long run goals of an optimal trend. This also explains in part the rationality of Japanese producers in building ahead of demand and concentrating on investment in technology-intensive industries. As Romer (1990) has pointed out, Japan now spends considerably more as a fraction of GDP on commercial non basic research that generates excludable benefits than does the United States and they are better off as a result.

To test the asymmetry of the two estimates, i.e., backward looking and forward looking we also applied the method of seemingly unrelated regression (SUR) for the labor demand equation but the estimated values of $\hat{\theta}_L = \Lambda_L/\Psi_L$ are found to be very close (see, e.g., Okamura (1994)). This suggests robustness of these estimates implying a strong asymmetry effect in the role of past history versus future expectations. A similar result is upheld for the capital demand equation, though the capital input is used here in a very aggregate form unlike labor. Thus the estimates from the forward-looking model show that future expectations play a very strong role in the long term calculations of Japanese producers. As Krugman (1991) has argued that in some dynamic environments future expectations may play a stronger role than the historical trends of the past and the case of Japan appears to fit in with this prognosis.

From an overall viewpoint our dynamic approach raises several important issues at an empirical level. First of all, the growth accounting approach in the Solow tradition does not distinguish between the roles played in the growth process by the past history and the future expectations. But in the new growth theory it is postulated that economies may reach different steady states because of differences in initial endowments and in the spillover of technical knowledge. Our empirical results provide some evidence in this direction. Secondly, our disequilibrium analysis provides an integrated macroeconomic analysis of adjustment lags and expectation formation. The paths of adjustment to the steady state are very dissimilar for labor and capital and even for the different components of labor. Export externality has a stronger impact on the speed of adjustment for the optimal labor demand compared to the demand for capital. In the heterogeneous labor model, the speed of adjustment increases two-fold for female labor and male skilled labor, when the export externality is allowed for in the optimal production frontier.

5.5 Concluding Remarks

The adjustment cost approach may be viewed as an adaptive method of learning by which the short run dynamics are sequentially updated. Although we have emphasized the linear quadratic model specification here, more nonlinear formulations are possible. Besides production and costs, other areas of economic applications include portfolio models of investment behavior, where the investor learns from the market and adjusts his portfolio over time. Here adjustment costs allow for risk sensitivity and the decision rules followed are in fact the so called Euler conditions of the variational calculus. Estimation of Euler equations has featured prominently in recent economic literature.

Table 1

Estimates of the production function with heterogeneous labor
(for different orders of the error process)

	$\Delta \ln Y_t$	Intercept	$\Delta \ln K$	$\Delta \ln F$	$\Delta \ln U$	$\Delta \ln S$	$\Delta \ln V$	R^2	DW
I.	AR(1)	-.025	.789	.194	.146	.040	-	.5	2.16
	t-value	(-.93)	(2.49)	(.67)	(-.68)	(.23)		0	
	AR(2)	-.011	.531	-.00	-.065	.162	-	.4	2.00
	t-value	(-.31)	(1.23)	(-.01)	(-.31)	(.68)		9	
	AR(3)	-.014	.618	.137	-.063	.065	-	.4	2.07
	t-value	(-.49)	(1.68)	(.38)	(-.27)	(.27)		6	
II.	AR(1)	-.023	.727	.222	.179	.055	.032	.5	2.17
	t-value	(-.82)	(1.90)	(.71)	(-.79)	(.31)	(.29)	0	
	AR(2)	-.010	.485	.016	-.076	.164	.026	.4	2.00
	t-value	(-.27)	(1.00)	(.04)	(-.35)	(.67)	(.24)	9	
	AR(3)	-.014	.588	.178	-.106	.071	.012	.4	2.03
	t-value	(-.45)	(1.45)	(.47)	(-.43)	(.29)	(.11)	6	

Notes:

1. Y_t denotes real GNP and V_t is exports in real terms.
2. DW statistics are all significant at 5% level.

Table 2

Solow residuals adjusted for exports in the heterogeneous labor model.

Year	Adjusted residual
1967	0.0242
1968	0.0329
1969	-0.0074
1970	0.0044
1971	-0.0263
1972	0.0154
1973	-0.0243
1974	-0.0422
1975	-0.0546
1976	-0.0052
1977	0.0024
1978	0.0783
1979	-0.0495
1980	0.0106
1981	0.0013
1982	-0.0073
1983	-0.0090
1984	0.0158
1985	-0.0318
1986	-0.0086
1987	-0.0061
1988	0.0153
1989	0.0309
1990	0.0048

1. Estimate of the adjusted residual ($\Delta \ln R_t$)

   mean: 0.0004
   maximum: 0.0241
   minimum: -0.0337

2. Correlation of

   $\Delta \ln R_t$ and ΔY_t: 0.4607
   $\Delta \ln R_{t-1}$ and ΔY_t: 0.6204
   $\Delta \ln R_{t-2}$ and ΔY_t: 0.1103

Table 3

Estimates of the speed of adjustment parameters (ϕ_i) and the weight ratio $\theta_i = \Lambda_i / \Psi_i$

Output ($\ln Y_t$)	Input ($\ln X_{it}$)	ϕ_i	θ_i
GNP (without exports)	*K*	0.702*	1.659
	F	0.335**	0.169
	U	1.071*	-16.42
	S	0.418*	0.299
GNP (with exports)	*K*	0.328*	0.162
	F	0.689*	1.524
	U	1.027*	-39.61
	S	0.826*	3.921

Notes:

1. One and two asterisks denote significant values at 5 and 10% levels of t test respectively.
2. Labor is shown here separately for skilled S and less skilled L man-hours and female F labor.
3. The stable characteristic root of the input demand equations can be derived from column ϕ_i by using the definition $\mu_i = 1 - \phi_i$. Note that except unskilled labor, the characteristic roots are all positive and less than unity.

Table 4

Two-step estimation of the labor demand equation in the two-input model (backward looking model)

First step

Dep.var.	Intercept	$\ln L_{t-1}$	$\ln Y_{t-1}$	$\ln RW_{t-1}$	$\ln V_t$	$\hat{\theta}_L$	DW	R^2
Z_t^L	-258.27	-	-3.545	-5.496	-	-	1.557	.608
	(0.49)							
Z_t^L	-103.75	-	-9.487	-1.527	3.686	-	1.749	.655
	(19.72)		(3.397)	(1.694)	(1.655)			

Second step

Dep.var.	Intercept	$\ln L_{t-1}$	$\ln Y_{t-1}$	$\ln RW_{t-1}$	$\ln V_t$	$\hat{\theta}_L$	DW	R^2
$\Delta \ln L_t$	17.913	-.9783	.186	.438	-	13.94	1.425	.552
	(3.530)	(.789)	(.065)	(.149)				
$\Delta \ln L_t$	8.931	-.898	1.859	1.973*	-.859	7.87	1.761	.809
	(2.409)	(.132)	(.591)	(.614)	(.139)			

Notes:

1. Estimated standard errors are in parentheses.
2. The symbol * indicates that instead of $\ln RW_{t-1}$ a proxy variable $\ln Y_{t-3}$ is used here as an explanatory variable since it had a more significant explanatory power.
3. V_t is used for real exports deflated by the export price index.
4. Here AR(3) processes are used to estimate the target demand variable d_t^L, since they provided the best value of the DW statistics when compared with AR(1) and AR(2) approximations.

Table 5.

Estimates of the externality effects of exports for the NICs in Asia

Country	Sample Period	a_0	a_1	a_2	b	R^2	ΔR^2
Korea	1967-86	-350.4	0.13*	0.03*	0.42*	0.657	0.089
Taiwan	1967-86	6070.9	0.12*	74.01	0.65*	0.936	0.346
Japan	1961-87	-7794.1	0.61*	0.12*	0.12	0.690	0.360

Notes:

1. Single asterisk denotes significance at 1% level of t-test.
2. ΔR^2 measures the increase in R^2 value when Δ exports is included as an explanatory variable.

References

Callen, T.S., Hall, S. and Henry, S. (1990), "Manufacturing stocks: Expectations, Risk and Co-integration", *Economic Journal*, vol. 100, pages 756-772.

Delong, J. and Summers, L.H., (1991), Equipment investment and economic growth, *Quarterly Journal of Economics* 106, pages 445-490.

Farrell, M.J., (1957), "The measurement of Production Efficiency," *Journal of the Royal Statistical Society*, Series A, vol. 120, pp.253-290.

Gregory, A., Pagan, A. and Smith, G., (1993), "Estimating Linear Quadratic Models with Integrated Processes." P. Phillips ed., *Models, Methods and Applications in Econometrics*, Cambridge: Blackwell Publishers.

Hall, R.E. (1988), The Relation between Price and Marginal Cost in U.S. Industry, *Journal of Political Economy*, 96(3), pp. 921-947.

Hall, R.E. (1990), Invariance Properties of Solow's Productivity Residual, in *Growth, Productivity and Unemployment*, MIT Press: Cambridge, Massachusetts.

Hay, G.A. and C.C. Holt (1975), A general solution for linear decision rules, *Econometrica*, Vol. 43, pages 231-260.

Kennan, J. (1979). "The Method of Partial Adjustment Models With Rational Expectations." *Econometrica*, vol. 47, pages 1441-1457.

Krugman, P. (1991), History versus Expectations, *Quarterly Journal of Economics*, 106(3); pages 651-667.

Kydland, F.E. and Prescott, E.C (1977), Rules rather than discretion: the inconsistency of optimal plans, *Journal of Political Economy*, Vol. 85, No. 3, pp. 120-135.

Lucas, R.E. (1993), Making a miracle, *Ecoometrica*, Vol. 61, pages 251-272.

Mankiew, N., Romer, D. and D. Weil (1992), "A contribution to the Empirics of Economic Growth", *Quarterly Journal of Economics*, 107, (2); pages 407-438.

Morrison, C.J. and Berndt, E.R. (1981), Short-run Labor Productivity in a Dynamic Model, *Journal of Econometrics* 16, pages 339-365.

Norsworthy, J.R. and S.L. Jang (1992), *Empirical Measurement and Analysis of Productivity and Technological Change*, Amsterdam: North Holland.

Okamura, K. (1994), *Returns to Scale and Export External Effects*, Ph.D. Dissertation, University of California, Santa Barbara.

Pilat, D. (1993), Explaining Japan's postwar economic growth: the contribution of growth accounting, in *Explaining Economic Growth*, New York: Elsevier Publishers.

Romer, R. (1990), Endogenous technical change, *Journal of Political Economy*, 98(1); pages 71-103.

Sengupta, J.K. (1995), *Dynamics of Data Envelopment Analysis*, Boston: Kluwer Academic Publishers.

Sengupta, J.K. (1993), Growth in NICs in Asia: Some Tests of New Growth Theory, *Journal of Development Studies*, 29(2); pp. 45-60.

Sengupta, J.K. (1992), Adjustment costs and the production frontier analysis, *Economic Notes*, Vol. 21, pp. 316-329.

Simon, H.A., (1956), *Models of Man*, New York: Wiley.

Treadway, A.B. (1969), On rational entrepreneurial behavior and the demand for investment, *Review of Economic Studies*, Vol. 36, pages 227-240.

Young, A. (1991), Learning by doing and the dynamic effects of international trade, *Quarterly Journal of Economics*, 106; pages 369-406.

CHAPTER 6: Variable Structure Systems

6.0 Introduction

Since its introduction in the early sixties, optimal control has been used extensively to determine strategies in many areas of economics. Economic policies based on closed loop optimal control (CLOC) laws are the most powerful and stable, but yield a control vector with continuously changing components. Such policies work well in organizations or institutions with reliable information, with a centralized small group of decision makers, and with clear communication lines with the agents implementing the policies. The reality of many decentralized organizations is that poor communications and model inaccuracies are often the source of discrepancy between the expected and actual performance of their policies. The inaccuracies can be either structured uncertainties (inaccuracies in the coefficients estimated for the model) or unstructured (omission of relevant variables, lags, underestimation of the order of the model, and any unmodeled dynamics). Since model inaccuracies have strong adverse effects on system performance, it seems natural to implement a set of laws that are simple enough to allow for fast implementation, that are robust to model inaccuracies, and that still yield a near-optimal performance. Two complementary tools have been developed to cope with the degradation of system performance due to uncertainty; Adaptive control, which is introduced in the next chapter, and variable structure control.

The design of variable structure control (VSC) developed in the late fifties by a group of Russian mathematicians. The sustained interest in VSC is due to its three main attributes: (1) a fixed set of rules is established, and if certain conditions are met, the rules are not affected by changes in some of the model parameters, (2) the formulation allows for piecewise continuous control, which is particularly relevant for economic policies (where exceeding a target by 10% usually does not have the same consequences as being 10% short of the target), and (3) the main advantage of the VSC solution is that it is robust to misspecification of the model, and in some cases will lead to a stable steady state solution while the CLOC solution yields an unstable system. Because it is extremely robust, VSC can be used successfully to control an economic model in the presence of model uncertainties (errors always exist in practice as a result of imperfect estimates of the state or the system parameters).

6.1 The variable structure control formulation

6.1.1 Overview.

The initial spark of interest came from the invention of extremely fast electrical switches called high speed relays. In 1955, Tsypkin showed that control laws, traditionally implemented with the use of analog electrical devices, could also be implemented using high speed relays. Experiments with relays led to the discovery that VSC could synthesize systems which are insensitive to disturbances. The robustness properties of VSC were investigated by Emel'yanov (1959), who also defined the necessary and sufficient conditions for the existence of a sliding regime in the continuous time formulation. A rigorous mathematical definition of motion in a sliding regime was given by Filippov (1964), and a formal method to devise a control that would yield a specific system behavior was developed by Utkin (1971) and Drazenovic (1969). The first comprehensive books in English on VSC were written by Itkis (1976) and Utkin (1978). The development of boundary layer controllers by Corless and Leitman (1981) and Slotine and Sastry (1983), led to renewed interest in practical applications of VSC[1].

VSC is used for tracking problems; using piecewise continuous control action, the controller forces the variables of interest ("state" variables) to converge to and stay on a desired surface (or path). The effectiveness of the control action is usually measured by a function of the distance between the path followed by the state variables (the "state trajectories") and the desired surface. The surface, called the "switching" or "sliding" surface, separates the phase plane (the coordinate system whose axes are the state variables) in regions in which different control laws are applied. Under certain conditions, once the state trajectory hits the surface, it enters a "sliding mode", and remains on the switching surface. In the presence of disturbances, time delays, and other model imprecision, the state trajectory does not always remain on the surface. Nevertheless, even in the presence of arbitrary parameter inaccuracies, the state trajectory can be made to remain in a neighborhood of the surface using a different control law on either side (hence the name "variable structure").

Economic policies of the kind suggested by this approach have been frequently used. For example, the deficit reduction schedule of the Gramm-Rudman-Hollings (GRH) law defines a switching surface. In the context of the GRH law, an economic policy is best if it reduces the budget deficit according to the schedule. If the deficit exceeds what is mandated by the law, the control action is to implement an across-the-board cut in government expenditures to bring the deficit back on schedule. If the deficit is smaller than what specified by the schedule, the

1 The idea of using a boundary layer has been known in the Russian literature since the 1960s. Applications of the method have been rare because the VSC law can require frequent switching between the set of controls, and high frequency switches were not available. Recent progress in very fast switching devices has made the implementation of VSC laws possible and has renewed interest in the method. For a list of the references to representative selection of applications of VSC, see DeCarlo et Al.(1988), p230, p230. and for an application in economics, see Fanchon and Fadali (1994).

control action is to use a mix of monetary and fiscal policies to increase economic welfare.

A very interesting property of VSC is that during the sliding regime the system has invariance properties, in the sense that the state trajectory becomes independent of some of the parameters of the system. Consequently, once the state trajectory enters a sliding regime, the system can become extremely robust to misspecification. This property makes VSC particularly suitable to economic problems where the parameters of the underlying economic model lie within large confidence intervals, where econometric assumptions are violated, or where misspecification plagues the model.

The VSC methodology can be used with non linear systems of arbitrary order. However, since many economic models are formulated as a linear system of first order differential equations, we will concentrate our exposition of VSC methodology to first order linear systems. For a more general exposition, we refer the reader to Utkin's work (1978) and Slotine and Li (1991). Hence we only consider systems linear in the control and characterized by the state equation:

$$dx/dt = f(x,t) + B(x,t)u(x,t), \qquad (6.1.1)$$

where $x \in \Re^n$ is the state vector, $u \in \Re^m$ is the control vector, $m < n$, and the m-dimensional switching surface is defined by:

$$s(x,t) = 0 \qquad (6.1.2)$$

where $s(x,t) \in \Re^m$. The design of VSC involves two steps; the design of the switching surface, and the formulation of a control law that will ensure global asymptotic stability of the system.

6.1.2 Design of the sliding surface.

The sliding surface is designed such that the system has a desired response both on the surface and in its neighborhood. A sliding motion occurs on the surface if the directions of motion of the state trajectories on either side point towards the surface. In the multivariate case, sliding motion occurs on the intersection of m surfaces s_i (i=1 ..m) rather than on just one surface. Each surface s_i corresponds to one of the m components of the control vector, and the intersection of all the surfaces s_i define the multivariate sliding surface $s(x,t)$. Filippov (1964) has shown that a set of sufficient conditions for the existence of a sliding regime is that the relationships:

$$s_i(ds_i/dt) < 0, \qquad i = 1,2 \ldots ,m \qquad (6.1.3)$$

hold in a neighborhood of the sliding surface, where s_i is the ith component of the multivariate function $s(x,t)$. Hence in a neighborhood of the surface, we must have:

$$\lim_{s_i \to 0^+} \left(\frac{ds_i}{dt}\right) < 0 \quad \text{and} \quad \lim_{s_i \to 0^-} \left(\frac{ds_i}{dt}\right) < 0.$$

Equation (6.1.3) implies that in a neighborhood of the surface defined by $s_i=0$, the system has a degenerate Lyapunov function of the form $v(x,t) = \frac{1}{2}s_i^2$, and is therefore conditionally stable relative to the surface. A condition that not only guarantees the existence of a sliding regime, but also guarantees that the sliding regime will be attained in finite time is: $\frac{1}{2}s_i(ds_i/dt) \leq -\eta|s|$, where η is a strictly positive constant[2]. Condition (6.1.4) is central to the design of the control law; First, it implies that the state trajectories must point toward the sliding surface in a neighborhood of the surface. Second, it implies that the control law must be discontinuous as the state trajectory crosses the sliding surface. These two features are exploited to give the system the desired dynamic properties, at the cost of having a discontinuous control law. An enormous advantage of VSC is that the system can be made to slide along a surface that is known with no uncertainty (since it is designed by the controller), even with a very imperfect knowledge of the parameter estimates (or even of the functional form) of the controlled system. All that counts is that the state trajectories are made to point towards the surface. Of course, there will be a trade-off between the speed at which the state moves along the sliding surface and the cost of the control action.

Since any surface satisfying (6.1.3) will have a sliding regime, one can design a surface that will satisfy the equation and also generate a specific system behavior. Two method have gained wide acceptance; Filippov's method, and the method of *equivalent control* introduced by Drazenovic (1969) and Utkin (1971). The method of equivalent control is the simplest one, and uses the fact that while in a sliding mode and on the surface, equations $s_i(x,t)=0$ and $ds_i(x,t)/dt=0$ are both satisfied.

We can now proceed to build the sliding surface given an estimate of the system dynamics.

Consider the (possibly time varying) system with equation $dx/dt = f(x,t) + B(x,t)u$, where $x \in \Re^n$, and $u \in \Re^m$. Substitution of $f(x,t)+B(x,t)u$ for dx/dt in $ds(x,t)/dt = 0$ and solving for u, yields the equivalent control;

---

2 The relationship between $\eta|s|$ and the width of the boundary layer around the sliding surface within which the system will enter a sliding mode is analyzed at length in Slotine and Li (1991). In particular, if there are bounds on the possible values of the control, as is often the case in economics, the boundary layer can be the of limited width. This extra set of constraints can also limit severely the speed of convergence towards the surface.

$$u_{eq} = -\left(\frac{\partial s}{\partial x}B\right)^{-1}\left(\frac{\partial s}{\partial x}f + \frac{\partial s}{\partial t}\right) \tag{6.1.5}$$

where $[\partial s/\partial x\ B]$ is non singular. The equivalent control is the control that will maintain the state trajectory on the surface (i.e. which will maintain the sliding regime) provided the system dynamics are known exactly. In economics, the equivalent control is the policy that will keep the economic targets on their desired paths. It is important to note that the dimension of the sliding surface is not necessarily equal to the dimension of he state space. In effect, the sliding surface identifies which of the components of the state vector are important to the controller, and the "unimportant" components are left to assume any possible value. Hence, the system of order n is reduced to a system of order $(n\text{-}m)$.
Substitution of u_{eq} in the state equation yields the state trajectory described by

$$dx/dt = f - B\left(\frac{\partial s}{\partial x}B\right)^{-1}\left(\frac{\partial s}{\partial x}f + \frac{\partial s}{\partial t}\right) \tag{6.1.6}$$

where f, B and s are functions of x and t. Finally, if the controller wants a system with behavior described by $dx/dt=h(x,t)$, he must solve: $f - B[\partial s/\partial x\ B]^{-1}[\partial s/\partial x\ f + \partial s/\partial t]=h(x,t)$ to identify the surface $s(x,t)=0$ that will give the system the desired dynamics. Note that if the conditions of the implicit function theorem are satisfied on the switching surface, up to m of the state variables can be expressed as a function of the others. Their expressions can then be substituted in the system of state equations, reducing it to a set of at least $n\text{-}m$ differential equations, and making it invariant to fluctuations in some of the state parameters. We now illustrate this invariance property; Consider the linear time invariant system[3]: $dx/dt=Ax+Bu$. Application of the method of equivalent control with the surface $s=Cx$ yields the simple control law

$$u_{eq} = -(CB)^{-1}CAx, \tag{6.1.7}$$

where C is an m by n matrix (whose rows are the gradient vectors of the surface $s(x,t)=0$), and CB is assumed nonsingular. The system behavior during the sliding mode is given by

$$dx/dt = A_{eq}x \tag{6.1.8}$$

3 Although general non-linear switching surfaces are possible, linear ones are more prevalent in design. For a very clear example of the design of a non linear surface, see DeCarlo et Al. (1988).

where $A_{eq} = [I\text{-}B(CB)^{-1}C]A$. Note that since $[I\text{-}B(CB)^{-1}C]$ is an idempotent matrix, it is a projector, and the projection of the state space (the "range space") has a smaller dimension than the state space. If A is of full rank, we have; $\text{rank}[A_{eq}] = \text{rank}[I] - \text{rank}[B(CB)^{-1}C] = n - m$. Hence the equivalent control is unaffected by variations of m components of the state vector[4].

6.1.2 Design of the feedback structure.

The essence of VSC is to design a control law that will keep the system in a sliding mode along a desired surface. That is achieved as long as the sliding conditions (2.3) are met. Without model imperfections, and once on the desired surface, the state trajectory will remain on it when the equivalent control (u_{eq}) is applied. However, model imperfections can cause the state trajectory to deviate from the desired surface, and a piecewise continuous control law must be used to bring the state trajectory back on it. Therefore, the control law must be such that in a neighborhood of every sliding surface, the state trajectories point toward the surface as shown in Figure 1. To that effect, different controls (denoted by u_i^+ and u_i^-) are defined on each side of the surface, and the VSC law is of the form:

$$\begin{aligned} u_i &= u_{eq\,i}(x,t) && \text{when } s_i(x,t) = 0, \\ u_i &= u_i^+(x,t) && \text{when } s_i(x,t) > 0 \\ u_i &= u_i^-(x,t) && \text{when } s_i(x,t) < 0, \qquad i = 1 \ldots m. \end{aligned} \tag{6.1.9}$$

4 This invariance property is a by-product of equation (6.1.2): In a system with perfect information, no hysteresis, no delays in the control action, and no unmodeled dynamics, once the system enters a sliding mode, the surface equation (6.1.2) becomes an identity describing the state trajectory. Hence in an ideal situation, the identity can be used to eliminate m of the n components of the state vector. In practice, and especially in economics, it is rare to find m independent components of the state vector, and the reduction in the dimension of the system is likely to be less than m.

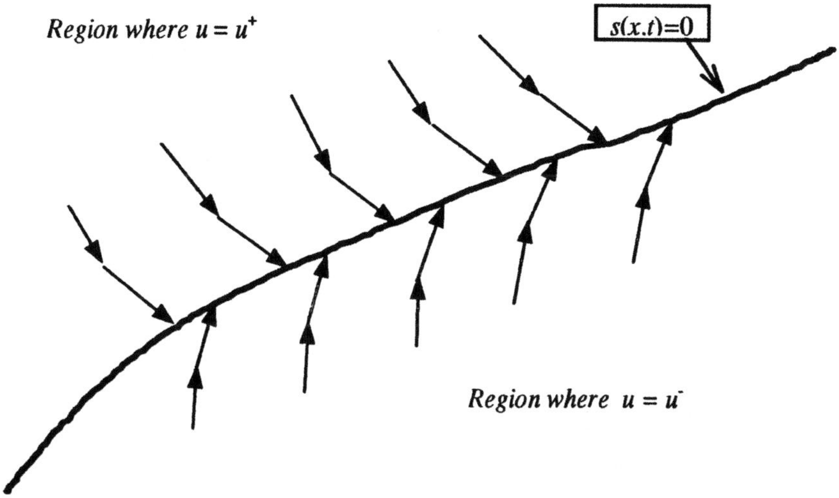

State trajectories around a sliding manifold

Figure 1

Model imperfections can still affect the position and bearing of the state trajectories in a neighborhood of a sliding surface, but once the system enters a sliding mode, it will remain in that mode as long as the affected trajectories still point toward the surface, and only the speed of sliding is changed (which explains the robustness of the method): If a control u_i yields a state trajectory which is nearly tangent to the sliding surface, the sliding speed is near its maximum. However, if the model is imperfectly specified or estimated, the actual state trajectory will be different from what was expected, and the chosen control law might be insufficient to maintain the sliding regime (figure 1a).

Conversely, if the trajectories follow the gradient vectors of $s(x,t)=0$, the system will enter a sliding mode, but the sliding motion will be null or very slow. In conclusion, under ideal conditions, the controller could force the system to follow any given path, but in practice, the controller must find a balance between sliding speed (i.e. speed of convergence) and robustnes to misspecification. Two elements have held back the use of VSC for economic applications: the "Chattering problem" and the lack of an objective function. In the following sections, we introduce modifications of the control law that makes VSC a suitable tool for economics.

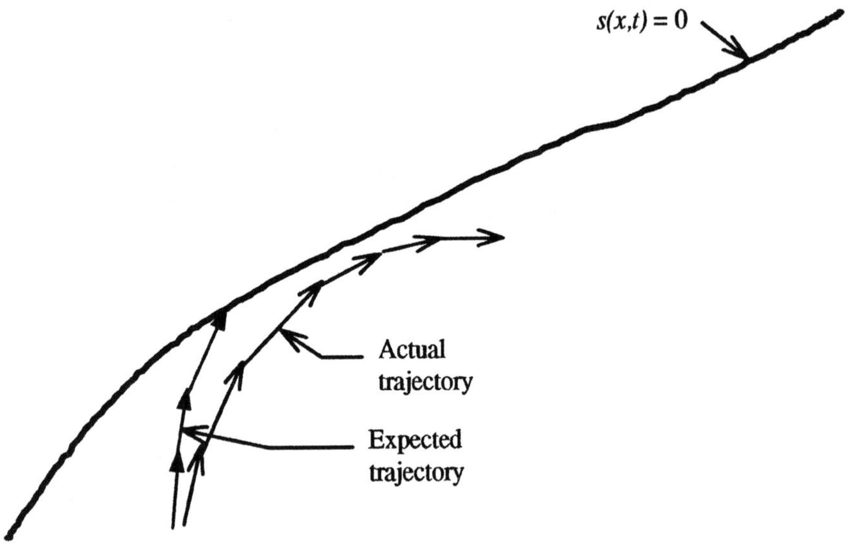

Actual versus expected state trajectories

Figure 1a

6.1.3 The Chattering problem.

In theory, it is possible to maintain a sliding mode using the control law (6.1.9), which assumes an immediate control action and perfect information about the state variables. In practice, model imperfections, lags and imperfect information will cause the state trajectory to deviate from the desired surface, and the components of the control vector will often switch between u_i^+ and u_i^-. The resulting frequent switching of control is called "chattering". Whether chattering is good or a bad is still debated today. There is growing evidence that under some circumstances, a policy with frequent control switches can outperform a policy with few or no switches. Several papers[5] provide an empirical proof of the benefits of a chattering policy.

---

5 Sasieni (1971) proposes a non linear model whose optimal solution is a chattering control. More recent models with a chattering control have focused on the non concavity of the objective function, or of the system dynamics. For example, see Mahajan and Muller (1986), or Mesak and

In practice, chattering control is possible only when the control action can be implemented instantaneously. The controller must be able to estimate the state of the system precisely, and with no delay, in order to be able to determine what is the proper control action. Such conditions might exist in firms with excellent communication channels, and with the ability to process the information needed very rapidly. Hence it is not surprising that the models where the optimal control is a chattering control are all from microeconomics and marketing. Because regional and national data are not instantly available, and because the political and monetary costs associated with switching policies are very high, chattering control is generally undesirable in macroeconomics.

There are several ways to reduce the chattering problem. The discontinuity of the control action can be smoothed by using integral control[6], or by defining a boundary layer around the surface[7]. In this chapter, we illustrate the second approach; Within an $e_{max,i}$-neighborhood of each sliding surface, i.e. when $|s_i| < e_{max,i}$, we use the equivalent control[8] defined by (6.1.5), and use the VSC control outside the boundary layer.

Even though the values $e_{max,i}$ (one for each surface) called the maximum tolerable tracking error for the i-th surface, can be specified individually, they can also be specified by a more general relationship among them: $e_{max} = \psi(e_{max,1}, e_{max,2}, e_{max,m})$. The piecewise continuous control law is described by:

Darrat (1992). For an excellent summary of the state of research on "chattering" and "pulsing" policies in advertising, see Feichtinger, Hartl, and Sethi (1994).

6 Integral control can be used when the system is not subject to exceptionally large disturbances. The use of integral control to formulate economic policies was initially investigated by Mundell (1962), although not in the context of VSC.

7 The boundary layer approach, developed in the first part of the century, is characterized by a different control law being applied in a neighborhood of the surface. To avoid the discontinuity of the control when the state trajectory moves from one side of the surface to the other, the approach was later refined as explained in detail in Slotine and Sastry (1983) and Slotine and Li (1991): while the state trajectory is inside the boundary layer, the piecewise continuous control law is replaced by a continuous control law that provide a smooth transition from the control value on one side of the boundary layer to the control value on the other side.

8 Even though the equivalent control is the best control to maintain the sliding regime, any other control law that would keep the state trajectory within the boundary layer is acceptable. However, with a poor choice of control law, the trajectory will move out of the layer, where a possibly more costly control law will apply.

$$u = wu_{eq} + (1-w)u_N \tag{6.1.10}$$

where

$$\begin{aligned} w &= 1 \quad \text{if } |s| < e_{max}, \\ w &= 0 \quad \text{otherwise.} \end{aligned} \tag{6.1.11}$$

We can choose a linear control law of the form:

$$u_i = \sum_{j=1..m} k_{ij}x_j \qquad i=1..m \tag{6.1.12}$$

where the minimum gains k_{ij}^+ and k_{ij}^- are chosen such that the inequalities (6.1.4) are satisfied i.e. such that there is sufficient control to make the state trajectories point towards the surface. Hence to maintain a sliding regime, we must have:

$$k_{ij} = k_{ij}^+ < 0 \qquad \text{if } s_i x_j > 0, \tag{6.1.13}$$

$$k_{ij} = k_{ij}^- > 0 \qquad \text{if } s_i x_j < 0, \tag{6.1.14}$$

In other words, as long as the tracking error is small, the controller ignores it, and the control used in a neighborhood of the surface is the same as the control that would be used on the surface. If the tracking error exceeds the tolerable maximum, the components u_j^+ or u_j^- are used to bring the trajectory back towards the surface s_j. This approach can be viewed as a control problem with a boundary layer with constant width or thickness e_{max}. Note that there is a trade off between tracking performance and control activity: the smaller the maximum tolerable tracking error, the higher the control activity.

The Exchange Rate Mechanism of the European Monetary System provides a good example: participating countries are not required to intervene if the exchange rate between their currency and the European Currency Unit (ECU) fluctuates within certain limits (for example $\pm 2.5\%$ of the target exchange rate), and the country must intervene when the exchange rate exceeds the limits. In terms of VSC, the time varying switching surface is defined by the exchange rate between the currency and the ECU, and the maximum tolerable tracking error is $\pm 2.5\%$. If the exchange rate is within those limits, the equivalent control (for example a mix of fiscal and monetary policies) is used to maintain the parity. If a different control is used (e.g. political considerations might limit the choice of instruments), the exchange rate might drift beyond the limits allowed, and the country will then have to use a different control to bring the parity within the tolerable limits (devaluate or re-evaluate its currency), or increase the size of the maximum tolerable tracking error. A poor choice of policy is not the only way to leave the boundary layer:

external shocks to the economy, model and estimation inaccuracies, might also push the exchange rate beyond the allowed limits with the same consequences.

6.2 A VSC approximation of an optimal control law

Under ideal conditions (perfect information, no unmodeled dynamics, ...), the CLOC law defines the best policy to optimize an objective function. However, the reality of economics is that policies are often formulated using imperfect information, inaccurate modeling, and their performance can differ substantially from what is expected.

VSC does not optimize an objective function. However, VSC can be used to track the path of some variables, provided the sliding conditions hold on the desired path; Since policies are adopted for their expected performance and not their actual performance, we propose to first determine what would be the expected path of the variables under the CLOC law as identified at the time the policy is formulated, and then devise the VSC law that will keep the trajectories on or near the expected path.

The first step is to determine the dynamics of the system under the CLOC law. For a linear system, this boils down to identifying the eigenvalues of the closed-loop state feedback matrix[9] A_{cl}, which describes the path of the state variables under the CLOC law, (i.e. $dx/dt = A_{cl}x$).

In a second step, we design a sliding hyperplane such that a VSC law will yield a system with approximately the same dynamic properties the system is expected to have under a CLOC law. Under the CLOC law, the equation of the state trajectory can be expressed as:

$$x_i(t) = a_{i1}e^{\mu_1 t} + a_{i2}e^{\mu_2 t} + \dots + a_{in}e^{\mu_n t}, \qquad i=1\dots n. \qquad (6.2.1)$$

where the μ_j's are the eigenvalues of A_{cl}. After a short transient period, the components with large negative ("fast") eigenvalues become negligible, and the dynamic properties of the system are determined by the exponential functions with the smallest negative ("slow") eigenvalues. Hence, if the matrix C defining a switching hyperplane

$$s = Cx = 0 \qquad (6.2.2)$$

is chosen such that it has the same "slow" eigenvalues of A_{cl}, the state vector in the sliding mode will quickly follow nearly the same path under the CLOC and VSC laws. The initial behavioral discrepancy of the system under the two control laws depends upon the number of "fast" eigenvalues not considered in building the

---

9 If the state equation is given by $dx/dt=Ax+Bu$, and the integrand of the objective function is $x'Qx+u'Ru$, the CLOC law is given by $u=-R^{-1}B'Px$, and $A_{cl}=A-BR^{-1}B'P$, where P is the solution of the algebraic Riccati equation $A'P+PA-PBR^{-1}B'P+Q=0$.

surface. The greater the number, the greater the difference between the path of the state vector under the VSC and CLOC laws.
The heart of the process is then to identify how many eigenvalues to use in the design of the hypersurface. To ensure that the system remains controllable, their number must not exceed the dimension of the control vector applied. Hence at most m eigenvalues of A_{cl} can be selected. For a linear system, the design of the hypersurface is greatly simplified by the following two theorems:

Theorem 1 (Young-Kokotovic-Utkin):
For any linear system $dy/dt = Ay + Bu$, *where* $y \in R^n$, $u \in R^m$, *and* $m<n$, *if the pair* (A,B) *is controllable, and if there exists a matrix* M *such that:*

1/ $CM^{-1} = (C_1\ C_2)$, (6.2.3)
2/ $MB = (0\ B_2)'$, (6.2.4)
3/ B_2 *and* C_2 *are* $(m \times m)$ *and non singular,* (6.2.5)

then the transformation $x=My$ *reduces the system to:*

$$\begin{aligned} dx_1/dt &= A_{11}x_1 + A_{12}x_2 \\ dx_2/dt &= A_{12}x_1 + A_{22}x_2 + B_2u \end{aligned} \tag{6.2.6}$$

where $x=(x_1,x_2)'$, $x_1 \in R^{n-m}$, $x_2 \in R^m$, A_{ij} *are conformal matrices, and the pair* (A_{11},A_{12}) *is also controllable.*

Proof: The controllability of (A,B) implies that: rank $\begin{bmatrix} A_{11}-\alpha I & A_{12} & 0 \\ A_{21} & A_{22}-\alpha I & B_2 \end{bmatrix} =$ n. Since (6.2.5) guarantees that we have rank$[B_2] = m$, we also have rank$[A_{11}-\alpha I, A_{12}] = n - m$ for all complex α. Hence (A_{11},A_{12}) is controllable.

Theorem 2 (pole placement theorem[10]):
If the system (A,B) *is controllable, the eigenvalues of the system under a linear state feedback of the form* $u = w + Wx$ *can be arbitrarily assigned.*

The first theorem guarantees that after the linear transformation $x = My$, the reduced system is controllable provided the original system is controllable. The transformation splits the transformed state vector x in two parts (x_1 and x_2), where x_1 is independent of the control vector. After the change of variables of

---

10 The proof is standard in most linear control books. for example see C.T. Chen, p279-280.

theorem 1, the sliding surface defined by $Cy = 0$ becomes $C_1x_1+C_2x_2 = 0$. Hence the second component of the state vector can be expressed as a function of the first component while the system is in a sliding mode (i.e. x is on the surface and satisfies $C_1x_1+C_2x_2 = 0$):

$$x_2 = (-C_2^{-1}C_1)x_1. \tag{6.2.7}$$

The system (6.2.6) can be viewed as a control problem with state vector x_1 and control vector x_2. In the sliding mode, equation (6.2.6) becomes:

$$dx_1/dt = Kx_1, \tag{6.2.8}$$

where $K = [A_{11}-A_{12}C_2^{-1}C_1]$ is the linear state feedback matrix.
The second theorem guarantees that since (A_{11},A_{12}) is controllable we can find a matrix $-C_2^{-1}C_1$ such that the eigenvalues of K are the same as the m "slow" ones selected. Definition of the sliding surface is therefore reduced to solving a standard pole assignment problem, and the control must be such that (6.1.3) and (6.1.9) are satisfied to keep the system on the sliding surface (or (6.1.3) and (6.1.10) to stay within e_{max} of it). The following section gives an example of the method.

6.3 Example: The expectations-augmented Phillips relation

6.3.1 The model

We will demonstrate an application of the VSC approach using a very simple model[11] of the Phillips relation, where money wage demands are a function of the expected rate of inflation. The assumed relationship between the actual inflation rate and the unemployment rate is:

$$p = \alpha - T - \beta U + h\pi, \tag{6.3.1}$$

where p is the actual rate of inflation, T is the exogenous percentage increase in labor productivity, U is the unemployment rate, π is the expected rate of inflation, α, β and h are parameters. The dynamics of the model is provided by;

11 The model, introduced by M. Friedman in 1968, is described in detail in Chiang (1984), pages 535-540. The simplicity of the example was the main criteria for the choice of this model, and the authors do not make any judgment about the economic assumptions made in its formulation.

$$d\pi/dt = j(p-\pi) \quad (0 < j < 1). \tag{6.3.2}$$

Monetary policy is the only policy instrument of the model. The desired target is to reach a steady state where the actual unemployment rate is equal to the natural level of unemployment, and where both the expected and the real rate of inflation are equal to zero. The rate of change of unemployment is assumed proportional to the difference between the growth rate of nominal money supply (m), and of the actual rate of inflation;

$$dU/dt = -k(m - p), \qquad (k > 0). \tag{6.3.3}$$

Equations (6.3.1) to (6.3.3) constitute a closed model in the three variables π, p and U. A few manipulations lead to a differential equation describing the time path of expected inflation. First, substitute (6.3.1) into (6.3.2);

$$d\pi/dt = j(\alpha - T - \beta U) - j(1-h)\pi.$$

Differentiation of the equation above with respect to time yields

$$d^2\pi/dt^2 = jdU/dt - j(1-h)d\pi/dt,$$

and using (6.3.3) to eliminate p gives the model:

$$d^2\pi/dt^2 = - j\beta k\pi - [\beta k + j(1-h)]d\pi/dt + j\beta km \tag{6.3.4}$$

After the change of variables $x_1=\pi$, $x_2=d\pi/dt$, $u=m$, $a_1=-j\beta k$, $a_2=-\beta k-j(1-h)$, and $b=-a_1$, equation (6.3.4) can be written in the usual control notation:

$$dx/dt = Ax + Bu, \tag{6.3.5}$$

where

$$x = \begin{pmatrix} \pi \\ d\pi/dt \end{pmatrix}, \qquad A = \begin{bmatrix} 0 & 1 \\ a_1 & a_2 \end{bmatrix}, \qquad B = \begin{pmatrix} 0 \\ 1 \end{pmatrix}. \tag{6.3.6}$$

Since the VSC formulation does not include an objective function, we specify one to solve the optimal control problem, and use VSC to steer the state on a path that will give the system approximately the same dynamic properties it would have under the CLOC law[12]. In this example, we assume that the objective of the policy maker is to minimize the function

---

12 The path adopted here is defined with respect to the optimal trajectory, but the control problem is essentially the same as that of keeping the economy on an arbitrary path. The fact that an optimization exercise is carried out is

$$J = \int_{to}^{\infty} [x'Qx+u'Ru]\, dt, \qquad (6.3.7)$$

where Q and R are both positive definite diagonal matrices. The elements of the diagonal of Q represent the relative weights attached to the dual objective of reducing expected inflation and reducing the change in the expected rate of inflation. Since the ultimate objective is to reduce the expected rate of inflation to zero, a greater weight is assigned to the first diagonal element of Q. The element R determines how the control action influences the objective function, i.e. the cost of changing the growth rate of the real money supply. For this example, we arbitrarily set $Q = \text{diag}(1, 0.2)$, and $R = ½$.

The problem, as formulated, is a linear quadratic regulator problem, and is solved in three steps. First, we estimate the coefficients of the matrices identified above. Second, we design the CLOC and VSC laws for the system. Finally, we do a simulation with both the CLOC and the VSC laws, and compare the resulting state trajectories.

6.3.2 Estimation.

A discrete version of the state equation (6.3.4) was estimated using non seasonally adjusted monthly data starting in 1959. Changes in the producers price index and M1 were used as a proxy for expected inflation and for the money supply, respectively. Ordinary least square[13] was used to estimate equation (6.3.4), with the restriction that the coefficients a_1 and b have equal magnitude and opposite sign. The estimated equation is:

$$\underset{}{d^2\pi/dt^2} = \underset{(.047)}{-.284\,\pi} + \underset{(.051)}{1.644\, d\pi/dt} + \underset{(.047)}{.284\, m} \qquad (6.3.8)$$

where the numbers in parenthesis are standard errors. All the regression parameters are significant at the .01% level, and the adjusted multiple coefficient of determination was equal to 0.7746.

not important because the parameters of the optimand are essentially arbitrary. Examples of this type of control exercise are Currie and Wren-Lewis (1990) and Weale et Al. ((1989), while a basic reference is Phillips (1954).

13 OLS is used to demonstrate that VSC design works even when the choice of estimation technique is inappropriate or very imprecise. We elaborate on this point in section 6.3.

6.3.3 Simulation.

The CLOC solution of the linear quadratic regulator problem defined by (6.3.5) and (6.3.7) is given by $u_\theta(t) = -K_\theta(t)x(t)$, where θ is a real and non negative coefficient of relative stability[14], $K_\theta = R^{-1}B'P_\theta x(t)$, and P_θ is the unique positive semidefinite solution of the algebraic Riccati equation (ARE):

$$(A+\theta I_n)'P_\theta + P_\theta'(A+\theta I_n) - P_\theta BR^{-1}B'P_\theta + Q = 0.$$

To the CLOC solution corresponds the closed loop state feedback matrix: $A_{cl\theta} = (A+\theta I_n-BK_\theta)$, which completely describes the optimal path of the state variables from given initial conditions. With the parameter values estimated above, we have:

$$K_\theta=\begin{pmatrix} 9.478 \\ 18.68 \end{pmatrix}, \qquad \text{and} \qquad A_{cl\theta}=\begin{bmatrix} 0 & 1 \\ -2.975 & -3.662 \end{bmatrix} \tag{6.3.9}$$

The eigenvalues of the matrix $A_{cl\theta}$ are $\mu_1 = -2.445$ and $\mu_2 = -1.216$. Since both are negative, the system is stable under the CLOC law. The time paths of the two components of x are shown in Figure 2.

The VSC solution is derived with the two steps described in section 6.1. Of the two eigenvalues of $A_{cl\theta}$, $\exp[-\mu_1 t]$ will converge the fastest to zero. Since we use a univariate control, only one "slow" eigenvalue can be selected (μ_2), and the switching surface is a line in the x_1x_2-plane. Note that since the system is already in the form (6.2.6), there is no need to find a transformation of the variables such that (6.2.3-5) are satisfied. Hence, in the notation of section 6.1 we have: $A_{11}=0$, $A_{12}=1$, $A_{21}=a_1$, $A_{22}=a_2$, and $B_2=b$. The transformation matrix M is the 2x2 identity matrix, and the matrix K of (6.2.8) is equal to $C_2^{-1}C_1$. Hence, $C_2^{-1}C_1=\mu_2$ is the solution of the pole assignment problem, and the slope of the switching surface is therefore chosen equal to μ_2. From equation (6.2.2), the switching surface has equation:

$$s = x_1 + 1.216x_2 = 0, \tag{6.3.10}$$

and the system behavior under the equivalent control is described by (6.1.8), where:

$$A_{eq}=\begin{bmatrix} 0 & 1 \\ 0 & 1.26 \end{bmatrix}$$

---

14 The coefficient of relative stability guarantees that all the optimal closed loop poles lie to the left of -q. For the definition, see chapter 2, and for an example, see Patel and Mundo (1982), pp213-215.

For the sake of clarity, we assume that the univariate control u_N defined by Equation (6.1.13) is a function of the first component of x (i.e. the change in money supply is only a function of the change in expected inflation). Hence $k_{11}^{+}=a$, $k_{11}^{-}=\beta$, $k_{12}^{+}=0$, $k_{12}^{-}=0$, and the control law (6.1.9) becomes:

$$u_N = \alpha x_1 \qquad \text{if } sx_1 > 0$$
$$u_N = \beta x_1 \qquad \text{if } sx_1 < 0. \qquad (6.3.11)$$

The sufficient conditions for the existence of a sliding mode (6.1.4) reduce to:

$$\alpha \leq (a_2 c - c^2 - a_1)/b \leq \beta, \qquad (6.3.12)$$

and substitution of the values of the parameters yield the restrictions:

$$\alpha < 0 \quad \text{and} \quad 2.83 < \beta. \qquad (6.3.13)$$

In contrast with the continuous time formulation described in section 6.1, the control law of this example is computed at discrete moments, and the same control is applied during the whole sampling period. Because the control is not adjusted as soon as the state trajectory reaches the sliding surface, a quasi-sliding mode (equivalent of the sliding mode for discrete systems) will appear where the trajectory frequently crosses the surface. The domain of the feedback gains that will make the state trajectory enter a quasi-sliding mode is a subset of the domain for the sliding mode in continuous time. To guarantee stability, condition (6.1.4) must be changed[15] to:

$$|s(x,t+\mathrm{h})| < |s(x,t)| \qquad (6.3.14)$$

where h is the sampling period, and (6.3.14) characterizes the quasi-sliding condition. The corresponding bounds defined in (6.3.13) must also be changed.

Finally, in this discrete formulation, the probability of $x(k)$ being a point on the switching surface is zero. If not modified, the VSC law will mandate very frequent switching between the two values u_i^{+} and u_i^{-}. The resulting chattering problem is greatly reduced if a law of the form (6.1.10-15) is applied[16]. If another control law (e.g. the optimal control law u_{opt}) is used in place of u_{eq} in (6.1.10), the trajectory will slowly drift to one of the edges of the boundary layer, being pushed back into the layer as soon as it crosses it. The chattering problem is therefore moved from the switching surface to the edge of the boundary layer, the severity of the problem

---

15 Sarpturk et Al. (1987) derive bounds for α and β to guarantee stability of the discrete system.

16 The relationship between the choice of the width of the boundary layer and the number of switches has been studied extensively. For example see Slotine (1984).

being affected by the width of the layer, by the discrepancy between the equivalent control law and the control law chosen, and by the unmodeled dynamics.

6.3.4 Comparison of the performances.

In all the simulations, expected inflation (x_1) and the change in expected inflation (x_2) were initially set at 20% and 10% respectively. The sampling period h was set at .01 month to approximate a continuous system (larger sampling periods produced similar results).

The first two simulations show the behavior of the system with various choices of control gains (α and β) in the VSC law (6.3.10). The width of the boundary layer (e_{max}) is set equal to zero to illustrate the chattering problem. In Figure 2, β is kept equal to 3 in all cases, and α ranges from -.6 to -3. These values of α and β are such that equation (6.3.14) is satisfied, hence once near the sliding surface, the VSC trajectories remain on or near the surface. In Figure 3, α is kept equal to -3 and β is varied. The state trajectory crosses the switching surface at the point K=(.1838,-.2357) where $s(K)$=.1029. With β=1.6, equation (6.3.14) is not satisfied beyond that point ($s(K$+h)=.1049), and the system becomes unstable since there is insufficient control to enter a quasi-sliding mode. With β=2 ($s(K$+h)=.1037) the same condition exists but the control action is sufficient to bring the trajectory back towards the surface, at a point where equation (6.3.14) is finally satisfied (i.e. where the state enters a quasi-sliding mode). With β=5, $s(K$+h)=.0933 and equation (6.3.14) is satisfied beyond K, when the trajectory enters the quasi-sliding mode immediately after crossing the surface.

The remarkable robustness of the VSC law is illustrated in Figures 4 and 5, which show the response of the system under the CLOC and VSC laws, when poor estimation technique, non linearities or other model imperfections yield incorrect values of the components of A and B of the state equation (6.3.5) (the VSC and CLOC laws were built from the estimated values a_1=-.284 and a_2=1.644). The simulations are done with e_{max}=.01 to reduce the chattering problem[17]. Figure 4 shows what the trajectories would be if a_2 was in fact equal to 1.6, and a_1 was in fact -.16, -.13, -.11 and -.10. Figure 5 shows the trajectories when a_1 is in fact equal to -.2, and a_2 is in fact equal to 2.0, 2.5, 3.0 and 3.5. In all the cases, the larger the error, the more unstable the system becomes under the CLOC law. The behavior of the system under the VSC law is a lot more robust to errors: After reaching the sliding surface, two of the four VSC trajectories do not enter a quasi-sliding mode in figure 4 and one in figure 5. In those three cases, an increase of the gain (α or β) is needed to satisfy equation (6.3.14). In figure 6, the trajectory x_{vs} describes the behavior of the system under the VSC law when no boundary layer

---

17 Even though the choice of e_{max} is arbitrary, e_{max}=.01 is chosen equal to a fraction of the standard deviation of the parameter estimates a_1 and a_2 of (6.3.8). With such a choice, the entire boundary layer is within the 95% confidence region for x_2.

is used (e_{max}=0 in (6.1.11)). The characteristic chattering problem of the VSC law is reduced by using the modified control law (6.1.10-15): x_{eq} is the state trajectory resulting from the use of the equivalent control u_{eq} (6.1.7) within a boundary layer of thickness e_{max}=05, and x_{op} is the state trajectory resulting from the use of the CLOC law only within the same boundary layer. For reference, we also plotted the state trajectory (labeled $x_{optimal}$) when the CLOC law is used both inside and outside of the boundary layer. The time path of x_1 under the various laws is given in figure 7. The great robustness of the VSC approach is exemplified in Figures 8 and 9, describing the behavior of the system under the same set of rules but when the estimated parameters are incorrect (In these simulations, a_1 is in fact equal to -.2, and a_2 is in fact equal to 2.5). The model imperfections push the state trajectories v_{vs} and v_{op} towards the edge of the boundary layer, while the x_{vs} trajectory still remains near the sliding surface. The optimal trajectory x_{opimal} does not follow the expected path and the maximum tolerable error e_{max} is exceeded. A similar difference in the behavior of the system under each of the two laws is obtained when a nonlinear system is approximated by a linear one[18].

---

18 The stability conditions for a general optimal control problem are usually difficult to establish. For a theorem specific to the linear quadratic regulator problem, see Patel and Mundo (1982), p 216-217.

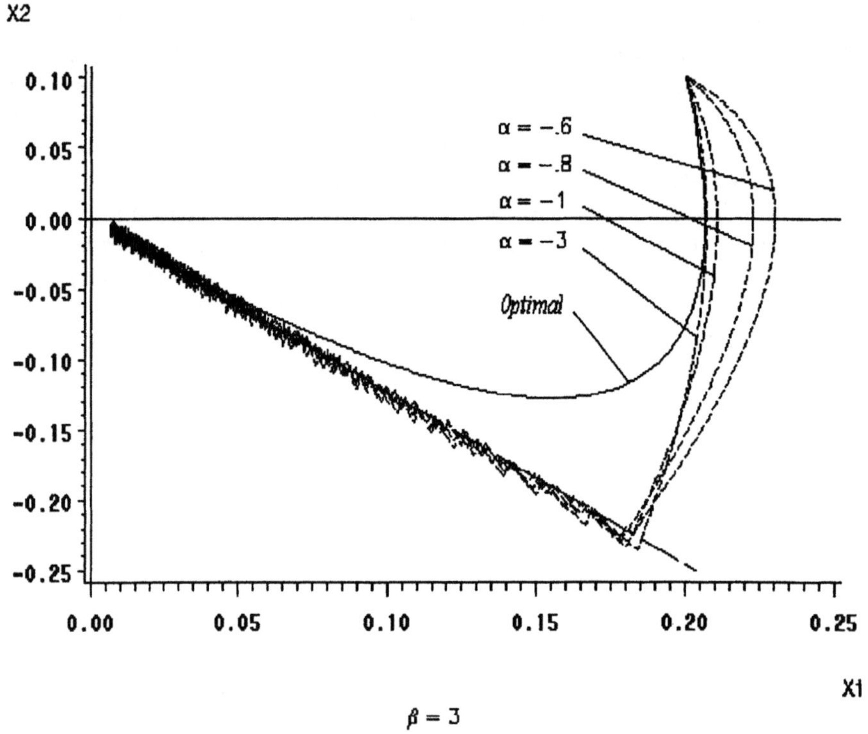

State trajectories under optimal control and VSC laws

Figure 2

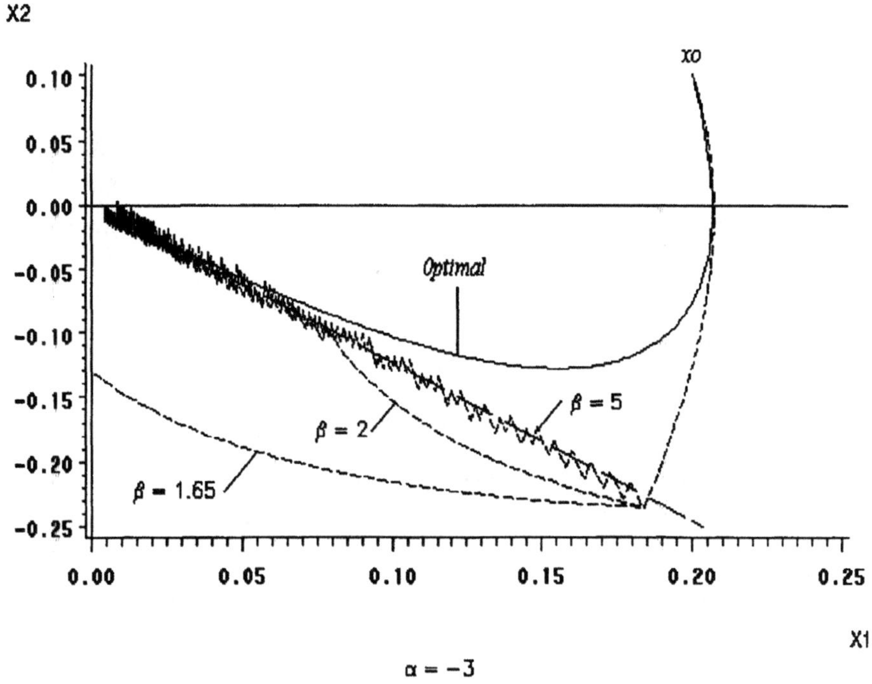

State trajectories under optimal control and VSC laws with decreasing gain

Figure 3

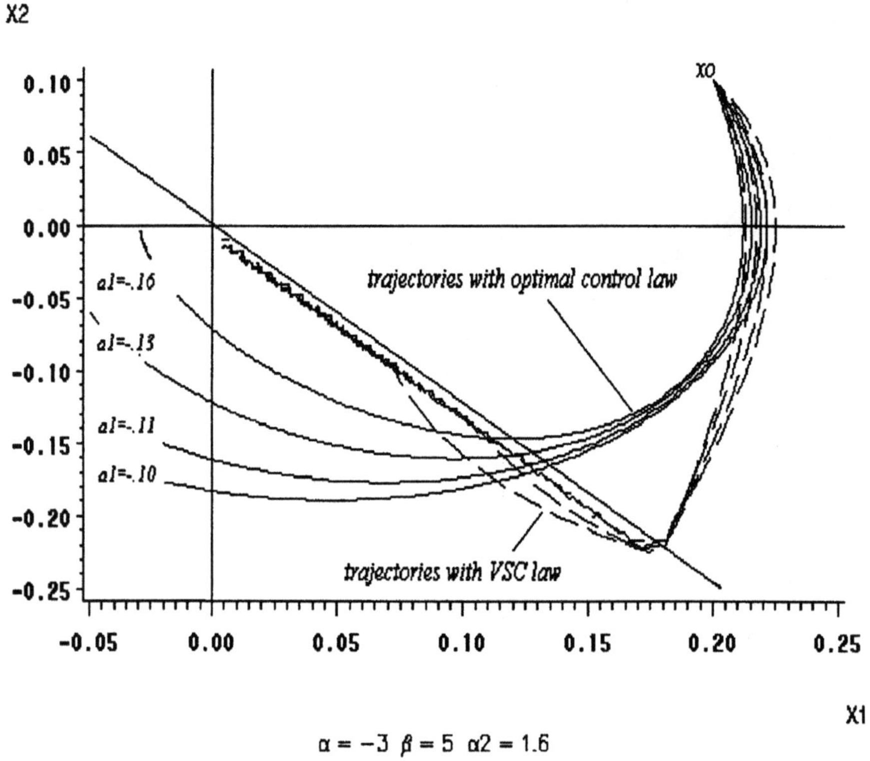

State trajectories under optimal control and VSC laws with a boundary layer and a misspecified model

Figure 4

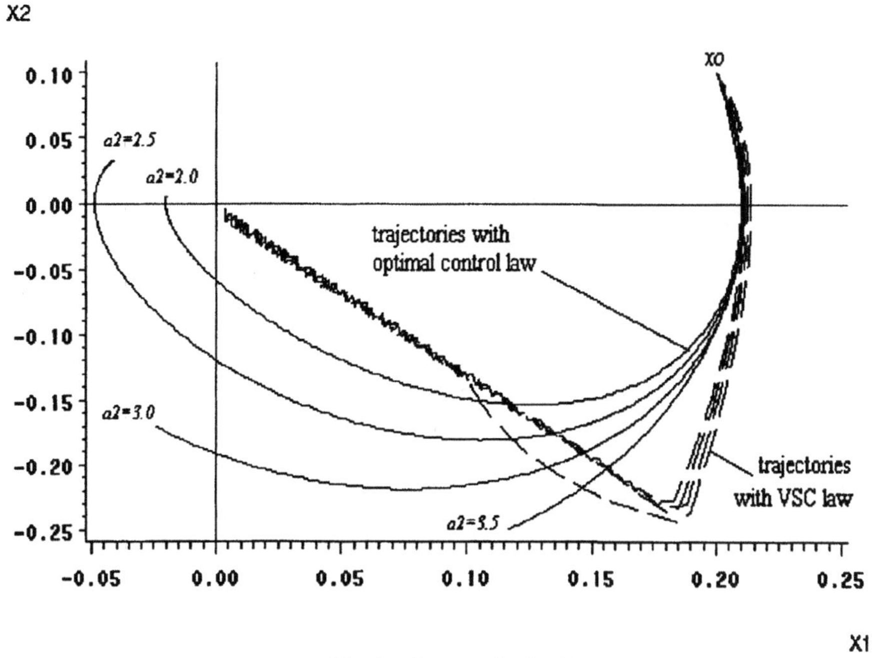

emax=.01 a1=-.2 $\alpha = -3$ $\beta = 5$

State trajectories under optimal control and VSC laws with no boundary layer and a misspecified model

Figure 5

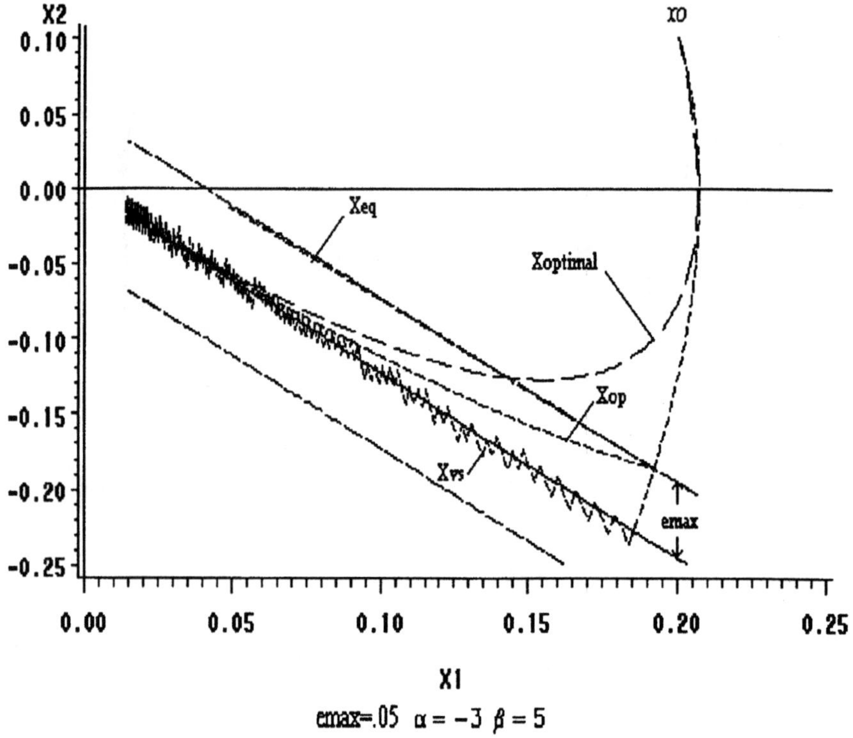

State trajectories under optimal, equivalent, and variable structure control laws

Figure 6

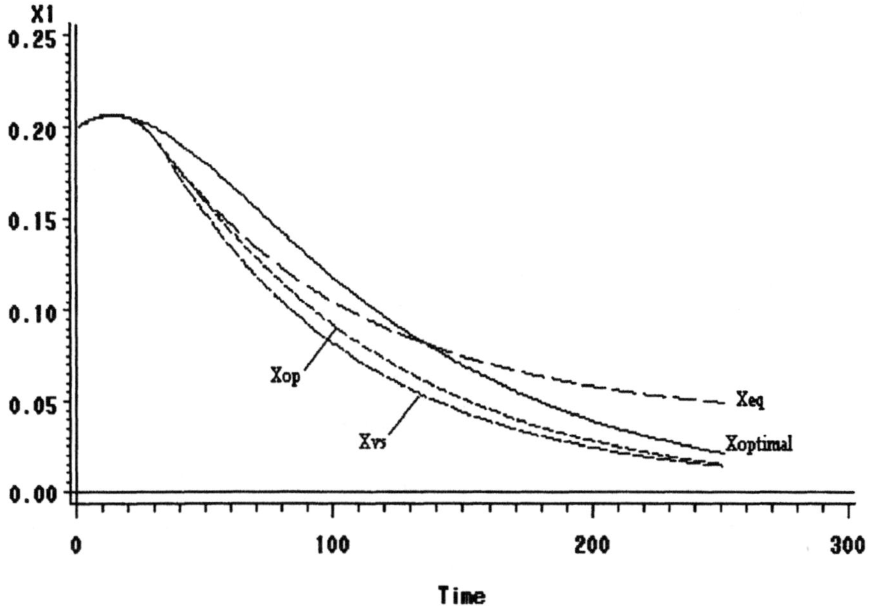

emax=.05 $\alpha = -3$ $\beta = 5$

Convergence to a steady state under optimal, equivalent, and variable structure control laws

Figure 7

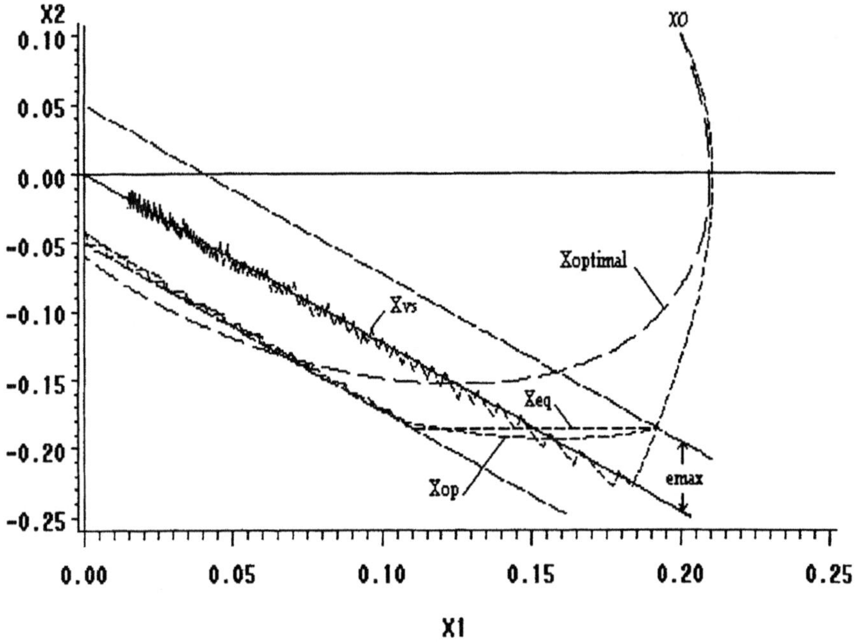

emax=.05 $\alpha = -3$ $\beta = 5$

State trajectories for a misspecified model under optimal, equivalent, and variable structure control laws, with a boundary layer

Figure 8

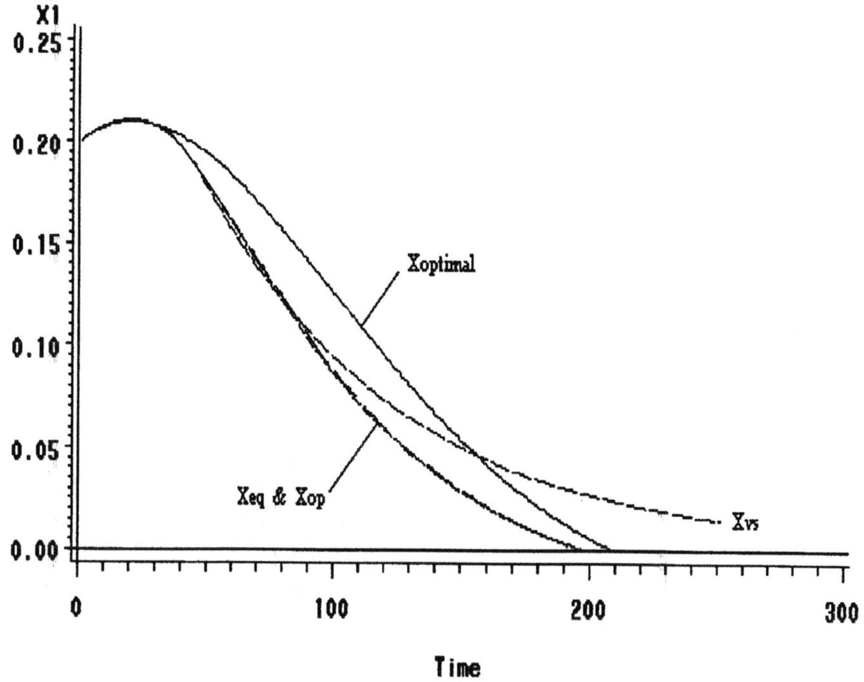

emax=.05 α = −3 β = 5

Convergence to a steady state for a misspecified model under optimal, equivalent, and variable structure control laws

Figure 9

6.4 Concluding remarks

The Variable Structure Control (VSC) approach provides a practical alternative to the more traditional design of policies based only on Closed Loop Optimal Control (CLOC). VSC laws are remarkably robust, which is why the method has numerous applications to large scale systems, uncertain systems, and non linear systems. However, implementation of a VSC law yields a potentially severe "chattering" problem (the frequent switch of control actions), which makes most macroeconomic policies based on the method unacceptable. The example introduces modifications to the usual VSC approach that produce a method more suitable for applications in economics and finance. Our primary contributions of are twofold;

First, we define a way to design a switching surface such that the system under the VSC law has dynamic properties similar to what they would be under the CLOC law. Our design allows for a reduction in the dimension of the control vector at the expense of tracking performance. The cost of a reduction in the number of instruments needed to achieve an economic goal is loss of optimality.
Second, we define a piecewise continuous control law where the "equivalent control" is applied within a boundary layer around the desired surface, and the VSC law outside the layer. Under such a law, the resulting state trajectories remain in a neighborhood of the desired sliding surface. The proposed law has the advantage of being very simple to implement because the control gains are constant within each region separated by the switching surface. Since the law is based on tracking a path rather than on optimizing a function, it also exceptionally robust to parameter variations in the input space (The image of the B matrix for linear input systems).
We use an example, with a discrete formulation of the technique, to illustrate the exceptional robustness of the proposed VSC law, and how a poor choice of law within the boundary layer will cause a frequent switching of control action along an edge of the layer. With increasing parameter errors, the performance of the system under the CLOC law deteriorates significantly, leading to instability while under the same conditions, the VSC law yields a stable system once it enters the quasi-sliding mode. The robustness of the method can be quite appealing to policy makers who could, within limits, make variables follow their expected path, whether or not the initial assumptions are correct.
VSC using sliding modes is a very powerful control method, but it is not perfect. It is a high gain control method whose performance can also be degraded in the presence of large parameter errors and unmodeled dynamics. In some cases, the control gain needed to satisfy the sliding conditions might be unacceptable to policy makers. A further line of research could investigate the trade off between constraints on the control action and tracking performance.

Even though the law proposed is not the "optimal" law, it duplicates the behavior of the system under a CLOC law and might well be the best practical one since the time needed to define and implement the "optimal" economic policy can easily render that policy less effective than a policy devised with the VSC approach. Policy makers can also choose any feasible policy within a boundary layer of the designed surface, with the guarantee that as long as the sliding conditions are satisfied, the economic system will behave as expected. This great flexibility for a choice of control within the boundary layer makes the approach particularly attractive to policy makers faced with a limited number of acceptable policies.

References

Chiang, A., (1984), *Fundamental Methods of Mathematical Economics*, McGraw Hill, 3rd. edition.

Corless, M. and Leitman, G. 1981 "Continuous State Feedback Guaranteeing Uniform Ultimate Boundedness for Uncertain Dynamic Systems," *IEEE Transactions on Automatic Control*, AC26:5 1139-1144.

Currie, D. and Wren-Lewis, 1990. "Extending the Target Zone Proposal for the G3." *Economic Journal*, Vol. 100, pp 105-123.

DeCarlo, R., Zak, S. And Matthews, G., (1988), "Variable Structure Control of Non-linear Multivariable Systems, *Proceedings of the IEEE*, 76(3), pages 212-232.

Drazenovic, B. 1969 "The Invariance Conditions In Variable Structure Systems" *Automatica*, 5:3 287-295.

El-Ghezawi, A Zinober and S. Billings, 1983 "Analysis and design of variable structure systems using a geometric approach", *International Journal of Control* 38:3 657-671.

Fanchon, P. and Fadali, S. (1994), "Design of Optimal Policies: A Variable Structure Control Approach", *Economics of Planning*, Vol. 27, pages 93-116.

Feichtinger, Hartl, and Sethi (1994), *Management Science*.

Filippov, A. (1964), "Differential Equations with Discontinuous Right Hand Sides," *Am. Math. Soc. Transl.*, 42 199-231.

Itkis, U., (1976), *Control Systems of Variable Structure*, Wiley, NY.

Mahajan, V. and Muller, E., (1986), "Advertising Pulsing Policies for Generating Awareness for New Products." *Marketing-Science*; 5(2), Spring 1986, pages 89-106.

Mesak, H., (1992). "An Aggregate Advertising Pulsing Model with Wearout Effects." *Marketing-Science*; 11(3), Summer 1992, pages 310-26.

Mundell, R.A., 1962 "The appropriate Use of Monetary and Fiscal Policy for Internal and External Stability." *IMF Staff Papers*. Vol 9, pp70-77.

Patel, R. and N. Mundo. (1982), *Multivariable System Theory and Design*, Perganon Press.

Phillips, A.W., (1954), "Stabilization Policy in a Closed Economy." *Economic Journal*, Vol. 64, pp 290-323.

Sarpturk, S., Y. Istefanopoulos and O. Kaynak, (1987) "On the Stability of Discrete-Time Sliding Mode Control Systems", *IEEE transactions on automatic control*, AC32:10 930-932.

Sasieni, M., (1989). "Optimal Advertising Strategies." *Marketing-Science*; 8(4), Fall 1989, pages 358-70.

Slotine, J.J., 1984 "Sliding Controller Design for Non-Linear Systems." *International Journal of Control*, 40:2 421-434.

Slotine, J. and Li, W., (1991) *Applied Non-Linear Control*, Prentice Hall.

Slotine, J. and Sastry, S., (1983) "Tracking control of Non-linear Systems Using Sliding Surfaces, With Application to Robot Manipulators," *International Journal of Control*, 38:2 465-492.

Utkin, V., (1971) "Equations of the Sliding Regime in Discontinuous Systems," *Automat. Remote Control I*, vol. 32, pages 1897-1907.

Utkin, V., (1972) "Equations of the Sliding Regime in Discontinuous Systems," *Automat. Remote Control II*, vol. 33, pages 211-218.

Utkin, V., (1978), *Sliding Modes and their Application in Variable Structure Systems*, Moscow: Mir publishers.

Utkin, V., and Young K., (1978) "Methods For Constructing Discontinuity Planes In Multidimensional Variable Structure Systems," *Automatika i Telemekhanika*, No.10, pp.72-77, October.

Weale, M, A. Blake, N. Christodoulakis, J. Meade and D. Vines, (1989), *Macroeconomic Policy: Inflation, Wealth and the Exchange Rate*. Unwin Hyman Ltd.

CHAPTER 7: Risk Sensitivity, Adjustment of Control and Estimation

7.0 Introduction

Risk sensitivity of optimal control rules in a dynamic model is a generic term for all types of responses to various shocks due to errors and uncertainties. The economic agent's response is to build elements of caution into his objective function e.g., the mean-variance approach in dynamic portfolio theory. The statistician's response to random shocks is to model the dynamic system such that the sample estimates of the random parameters are robust in some sense i.e. less sensitive to random fluctuations in equational errors. Two other situations are most important in applied economic models. One arises in dynamic or differential games, where one player's optimal control rules depend among other things on the control rules or strategies adopted by the other player or players. Mutual consistency of control or decision rules is thus an essential prerequisite e.g., Cournot-Nash equilibrium solutions explicitly allow for such mutual consistency. To estimate the optimal decision rule for each player in such a framework, one has to postulate the side condition specifying the second player's reaction function. Thus the method of maximizing the likelihood function subject to each player's reaction functions has to be adopted in the estimation process. Adjustment costs play a key role here. Finally, the economic agents tend to learn over time, when the observed data on the state variables become sequentially available. this learning process also exhibits a cautious risk attitude, which is strongly emphasized in current developments in control theory.

One important aspect of the estimation of a controlled system that is frequently omitted in the economic discussions of the optimal trajectory in a control model concerns the initial and terminal condition. In particular the latter, also called the transversality condition, which is one of the necessary conditions of an optimal control rule. We would discuss these and other related aspects of applied control theory. The basic motivation here is to illustrate some of the modern applied econometric techniques, their economic implications and also their problems.

7.1 Learning process in differential games

Consider a dynamic duopoly model where two players (i.e. suppliers) compete in a market e.g., the game theory models discussed in chapter 4, section 4.2.2 The world coffee market with brazil and Colombia as the dominant producers has been modeled in terms of Cournot-Nash type reaction functions. Consider the profit function for the i-th supplier:

$$\pi_i(t) = R_i(t) - C_i(t)$$

where:

$$R_i(t) = (\hat{p} + u)x_i(t) \qquad \text{(Revenue)}$$

$$C_i(t) = (c + v)x_i(t) \qquad \text{(Cost)}$$

$$\hat{p}(t+1) = \mathrm{E}[p(t+1) | S(t)] = a_0 - bS(t-1) + \beta S^2(t-1) \qquad \text{(Expected price)}$$

$$S(t) = \sum_{j=1}^{n} x_j(t) \qquad \text{(Aggregate supply)}$$

Here the simplifying assumptions are that the expected price and expected marginal costs are similar across different suppliers, u and v are random disturbance terms normally and independently distributed with zero means and finite variances σ^2_u and σ^2_v which may represent price and yield risks due to market and weather fluctuations. Each supplier has to forecast $\hat{p}(t+1)$ and then choose his own optimal supply $x_i^*(t)$ by maximizing the expected utility of profit $h_i = \mathrm{E}[U(\pi_i(t))]$ where the utility function is of the exponential form:

$$U = 1 - e^{-\alpha_i \pi_i(t)}, \quad \alpha_i > 0,$$

which has a constant rate of absolute risk aversion denoted by the positive parameter α_i. This yields the following maximand for each player:

$$\max h_i = [a - bS(t-1) + \beta S^2(t-1)]x_i(t) - \tfrac{1}{2}\alpha\sigma^2 x_i^2(t)$$

where $a = a_0 - c$, $\sigma^2 = \sigma^2_u + \sigma^2_v$. This yields the aggregate Cournot-Nash equilibrium solution which is consistent with each player's optimal decision rule:

$$S(t) = \theta_0 - \theta_1 S(t-1) + \theta_2 S^2(t-1)$$

where $\theta_0 = na(\alpha\sigma^2)^{-1}$, $\theta_1 = nb(\alpha\sigma^2)^{-1}$, $\theta_2 = n\beta(\alpha\sigma^2)^{-1}$. Two important implications of this solution may be noted. First, the nonlinear term $\theta_2 S^2(t-1)$ may contribute to oscillations due to complex roots of the associated quasi-polynomial. To see its implication, we may use the linearizing approximations $\dot{S}_i \cong S(t) - S(t-1)$ and $S^2(t-1) \cong \bar{S}^2 + [S(t-1) - \bar{S}](2\bar{S})$. Using the transformation $S(t) = e^{rt}$, this finally reduces to the following linear difference-differential equation (homogeneous part only)

$$\dot{z}(t) + rz(t) + \theta_5 z(t) = 0, \quad \theta_5 = \theta_3 e^{-\tau}.$$

This has the characteristic quasi-polynomial $\phi(\lambda) = \lambda + r + \theta_5 e^{-\lambda\tau}$, τ is the time lag, and $\phi(\lambda) = 0$ being the characteristic equation. Various regions of the solution space characterized by the real and complex values of λ may then be obtained. It can be shown that when $r > 0$ and $\theta_5 = 0$, the quasi polynomial has no roots with positive real parts and hence the system is asymptotically stable in this region. But if $\theta_5\tau > 1$, we obtain unstable oscillations. In this case the Cournot-Nash equilibrium solutions would not possess any stability property which may be desirable for the players. Secondly, the decline in the dominance of any one player (e.g., the decline of Brazil in the world coffee market) is most likely to reduce $\alpha\sigma^2$ or increase nb. This is then most likely to lead to an increase in the range of fluctuations of market supply. Note also that the time lags in the adjustment process characterized by the difference-differential equation above have also their adverse impact on the fluctuations of the state variable. Thus it is clear that an optimally controlled system in a differential game framework need not have the property of asymptotic stability without oscillations. This shows the need for a coordinated stabilization policy for the world commodity and other markets.

Critical to this coordination policy is the availability of market information to all the players, so that they can sequentially learn the adaptive strategies, which will reduce the overall market uncertainty and hence the market fluctuations. Learning about the parameters θ_1 and their own risk aversion may also help in the market stabilization.

7.2 Adjustment costs in portfolio models

For a long time the mean variance (MV) model has appeared to offer the most convenient and attractive approach to our understanding of the behavior of financial markets. However both theoretical and empirical tests over the last decade have not proved reassuring for this model. Recently two new approaches have been proposed, which provide alternative frameworks for appraising the MV efficiency model. One is the adjustment cost approach developed by Green (1988) and others who derive and empirically estimate the asset price expectation (APEX) equations by inverting a set of mean-variance-based asset demand functions. Green applied this technique for U.K. financial markets to test mean-variance efficiency and to estimate the parameters of the underlying utility function, notably the coefficient of relative risk aversion. Second, the standard MV efficiency frontier ignores the temporal variations in market volatility, although investors appear to be sensitive to risks due to such volatility. Recently Pindyck (1988) used quarterly data for the value-weighted New York and American Stock Exchange Index over 1963-64 to 1983-84 to estimate the effect of variance of real stock returns as a predictor of nonresidential fixed investment in U.S. He found the effect to be highly significant in a statistical sense, in the framework of multivariate regressions he conducted. Also one may note that if market volatility is measured by the conditional variance of returns, given the current state of information in the market, then its time-varying behavior may capture the temporal changes in market risks, which have been empirically investigated recently by a number of researchers. Recently Nelson (1991) has analyzed these time-varying conditional variance models, e.g., ARCH and generalized ARCH models and found for the weekly market indexes such as the S&P500 index and the value-weighted NYSE index that current returns and future market volatility are negatively correlated. Researchers beginning with Black (1976) have found this negative correlation between current returns and future volatility. Thus Backus and Gregory (1989) have given examples of equilibrium models where a regression of return on the conditional variance yields a negative coefficient. For the U.K. financial markets Green (1988,) applied his APEX equations over the monthly data for the period 1972-77 and found that the coefficient of relative risk aversion is significantly large and negative, where the relative risk aversion is defined in the standard Arrow-Pratt sense. These empirical estimates thus fail to support the major hypothesis of mean variance efficiency.

The object in this section is to introduce an adjustment cost approach in the MV efficiency model and apply it empirically to test the interrelationship between returns and market volatility. where the latter is measured by the conditional variances. This approach is more direct than Green's in two respects. First of all, it is based on an intertemporal cost function with return variances as components; thus it is more closely comparable to the minimum variance decision rules in asset allocation. Secondly, it allows the generalization that the impact of skewness and higher moments on the conditional return variances can be more directly evaluated.

7.2.1 Adjustment cost approach.

The adjustment cost approach proposed here assumes the convenient fiction of a representative investor who chooses a vector of portfolio shares x_t which minimizes a quadratic loss function subject to a lower bound on the expected return. The loss function is made up of the subjective variance of wealth $x_t'Vx_tw_t^2$ and the adjustment costs $(x_t - \mu x_{t-1})'A(x_t - \mu x_{t-1})w^2$, where w_t is wealth at market value (a scalar), $\mu = w_{t-1}/w_t$, V is an n by n subjective covariance matrix of asset returns and A is an n by n matrix of adjustment costs. The total adjustment costs here reflect the smoothness objective i.e., market value of asset holdings must not change too fast from one time point to another. For simplicity we assume $w_t = 1.0$ and $11 = 1.0$ and consider the simplest case first, where the investor minimizes one period loss function i.e.,

$$\text{Min } J = ½[x_t'Vx_t + (x_t - x_{t-1})'A(x_t - x_{t-1})] \tag{7.2.1}$$

$$\text{subject to } m'_{t+1}x_t = c_t$$

where m_{t+1} is the subjective expectation of the asset return vector and prime denotes a transpose. The inequality constraint $m'_{t+1}x_t > c_t$ is written here as an equality and it is assumed that the rational expectations hypothesis holds for all investors so that the subjective parameters (V, m_{t+1}) are estimated by their realized values in the market. The optimal portfolio x_t and the associated minimal variance $\sigma^2_t{}^*$ are then easily computed as follows:

$$x_t^* = [\alpha(V+A)]^{-1}[\alpha A x_{t-1} - \beta_{t-1} + c_t m_{t+1}] \tag{7.2.2}$$

$$\sigma^{2*} = \alpha c_t^2 - \alpha\beta_{t-1}c_t - \sigma^2_A$$

where $\alpha = [m'_{t+1}(V+A)^{-1}m_{t-1}]^{-1}$

$$\beta_{t-1} = m'_{t+1}(V+A)^{-1}Ax_{t-1}^*$$

$$\sigma^2_A = x_t^{*\prime}Ax_t^*.$$

Since the observed return variance σ^2_t in the market cannot be lower than the minimal variance $\sigma^2_t{}^*$ one obtains the variance function model:

$$\sigma^2_t = \alpha c_t^2 - \alpha\beta_{t-1}c_t - \sigma^2_A + u\,, \qquad u \geq 0 \tag{7.2.3}$$

In the steady state when x_t^* takes the equilibrium values x^*, the variance function model reduces to

$$\sigma^2 = \alpha c^2 - \alpha\beta c - \sigma^2_A + u\,, \qquad u \geq 0 \tag{7.2.4}$$

Since the expected value of u denoted by $E[u] = \bar{u}$ is positive, the method of ordinary least squares (OLS) is not applicable. One could however rewrite the right-hand side of the variance function (7.2.4) as:

$$\sigma^2 = v - \alpha\beta c - \alpha c^2 + \varepsilon\,, \tag{7.2.5}$$

where $v = \bar{u} - \sigma^2_A$, $\varepsilon = u - \bar{u}$, $E[\varepsilon] = 0$, and then the method of corrected ordinary least squares (COLS) developed by Richmond (1974) becomes directly applicable. By this method the parameters α, β, and v may be consistently estimated by regressing σ^2 on c.

The specification of the variance function models above has several interesting implications. First of all, if β or, β_{t-1} is positive, then the mean and variance of returns would be negatively correlated. This would then explain the puzzle empirically observed by Green for U.K. financial markets i.e., a highly significant negative value for the Arrow-Pratt measure of relative risk aversion. He concluded that a negative risk aversion is inconsistent with mean variance efficiency, since it implies either that agents are minimizing not maximizing utility, or that agents are risk seekers. Secondly, if the adjustment cost parameter matrix A is proportional to the variance-covariance matrix i.e., $A = kV$, $k>0$, then both β

and β_{t-1} would be positive and hence an increase in mean c_t would reduce the observed variance σ^2_t. Thus it is more likely that the mean and variance would be negatively correlated, although in general, β_t or, β may be negative at times. Thirdly, Nelson in his recent analysis modeled excess returns R_t as

$$R_t = a + bR_{t-1} + c\sigma^2_t + e_t \qquad (7.2.6)$$

where the conditional mean and variance of the error term e_t at time t are zero and σ^2_t respectively. On using daily returns data for the value-weighted market index from the CRSP tapes for July 1962 through December 1987 the estimated the parameters of the linear model (7.2.6) and found the coefficient c to be negative. Note that the formulation (7.2.1) which yields (7.2.3) and (7.2.5) is derived directly from the optimization objective implicit in the MV theory. Finally, one notes that the adjustment cost specification adopted here is of a quadratic form, since it is more tractable than a cubic form which allows for the asymmetry. However it is clear from the equation (7.2.6) that the marginal effect of σ^2_A on the market variance σ^2 is ordinarily negative, although the effect $\partial\sigma^2/\partial v$ may be positive in some domain if $\bar{u} > \sigma^2_A$. Thus the adjustment costs allow the flexibility in reducing or augmenting the observed variance in the series of market returns. For empirical data, this is an important feature that can be easily tested in a statistical sense. Furthermore, the variance function model (7.2.5) may be easily generalized to include a cubic regressor on the right hand side. Thus we may decompose the time series of returns Y_t into its conditional mean and conditional variance:

$$Y_t = E_{t-1}[Y_t] + e_t; \qquad \sigma^2_t = E_{t-1}[e_t^2]$$

where $E_{t-1}[Y_t]$ is the conditional mean and σ^2_t the conditional variance in period t, both depending on the information set I_t available up to period t-1. Let μ_{t-1} denote the conditional mean $E_{t-1}[Y_t]$. Then the variance function model may be expressed as

$$\sigma_t^2 = \gamma_0 + \sum_{i=1}^{4} \gamma_i (c_t - \mu_{t-1})^i + v \qquad (7.2.7)$$

where μ_{t-1} may be viewed as the target level of mean return. Asymmetry is now captured by the opposite reactions from the two events: $c_t > \mu_{t-1}$ and $c_t < \mu_{t-1}$. For example let i=3 and the regression coefficient γ_3 be negative. Then the skewness term in a conditional sense would augment the variance term σ^2_t if $c_t < \mu_{t-1}$ but the opposite reaction holds for $c_t > \mu_{t-1}$. This asymmetry hypothesis can be easily tested over the time series data on returns.

Here one needs to point out that the information set I_t that is used in the Arch model to define conditional means and variances may be interpreted more generally as a partition in the space of the distribution of returns. This partitioning introduces asymmetry in the sense that for any fixed c_t, the investor's reaction to the event $Y_t > c_t$, i.e., good news may be very different from that of $Y_t < c_t$, i.e., bad news. The standard Markowitz-type mean variance frontier characterizes the variance function $\text{var}(y_t|I_t)$ as a function of c_t alone, when the level of c_t is increased. But this characterization ignores the asymmetry in reaction. Here one can utilize an important theorem proved by Karlin (1982) which says that if the probability density $f(y)$ of Y_t is log-concave, i.e., $f(y) = e^{-\phi(y)}$, $\phi(y)$ being convex in y, then the conditional variance σ^2_t is strictly decreasing in c_t as c_t is increased in the positive domain. The class of log-concave densities is very wide, e.g., it includes normal, gamma, double exponential and all Polya frequency densities. This theorem can be applied empirically to market returns data by assuming that the entire market is divided into two groups of investors: the active group seeking $y_t > c_t$ and the remaining group which is passive, i.e., $y_t < c_t$. Whenever the active group is dominant in the overall market, the conditional variance $\sigma^2_t = \text{var}(y_t| \, y_t > c_t)$ would display an intertemporal behavior declining with the mean. In the opposite case when the passive investors dominate the market, e.g., in a bear market the conditional variance would tend to be increasing in relation to the mean level as c_t is increased. In our empirical applications we test the existence of the bull market asymmetric behavior for two sets of monthly data on market returns, i.e., the period January 1975 through December 1979 and the period August 1982 through December 1991. These two periods are characterized as bullish markets by Weissenberger Reports which are much relied on by the investor circles. Since the asymmetry (e.g., 'good news', 'bad news' effects) is likely to be more pervasive among the active and optimistic investors, these periods provide a more suitable framework for the empirical tests.

Next one may consider an intertemporal version of the quadratic optimization model (7.2.1), where the objective function is

$$\min J = \tfrac{1}{2}\sum_{t=1}^{\infty} \rho^t \left[x_t' V x_t + \left(x_t - x_{t-1}\right)' A\left(x_t - x_{t-1}\right)\right]$$

and ρ is the constant discount rate $0<\rho<1$. The optimal variance function model (7.2.3) and its steady state now appear as follows:

$$\begin{aligned} \sigma^2_t &= ac_t^2 - ab_tc_t + d_t + \sigma^2_{A(\rho)} + u, \qquad && u \geq 0 \\ \sigma^2 &= ac^2 - abc + d + v, && v = \bar{u} - \sigma^2_{A(\rho)} \end{aligned} \tag{7.2.8}$$

where

$$a = \{m'_{t+1}(V+(1+\rho)A)m_{t+1}\}^{-1}$$

$$b_t = \rho m'_{t+1} S A x_{t+1}^* + m'_{t+1} S A x_{t-1}^*$$

$$d_t = x'_t S A x_{t-1}^* + \rho x'_{t+1} A x_{t-1}^*$$

$$S = [(V+(1+\rho)A]^{-1}, \qquad \bar{u} = E[u]>0$$

$$\sigma^2_{A(\rho)} = x_t^{*\prime}(1-\rho)Ax_t^*$$

Clearly the structure of the solutions in the intertemporal and one-period optimization cases is very similar, except that the exogenous discount rate ρ is most important here influencing such terms such as $\sigma^2_{A(\rho)}$ and S which reflect elements of cautionary behavior. Since one has $\sigma^2_{A(\rho)} > \sigma^2_A$ for any positive value of ρ, therefore the term v may tend to be negative for large ρ, thus reducing the observed estimate of the return variance σ^2_t. Also, compared to the regression coefficients α, β of (7.2.4) the new coefficients a, b of (7.2.8) are higher i.e., $\alpha < a$, $\beta < b$ for any positive discount factor. These imply that the variance function models (7.2.4) and (7.2.8) may differ in the empirical phase due to the prevailing level of risk aversion implicit in the positive value of ρ which reflects impatience.

7.2.2 Risk sensitive frontier.

A basic objection to the adjustment cost formulations above is that it assumes like the MV efficiency theory that the utility function is either. quadratic, or exponential with a random normal distribution for portfolio returns. For a more general framework one needs an extension of this quadratic approach. The extension ought to incorporate risk sensitivity by the investors in the capital market.

Let w_0 denote initial wealth measured in monetary units and assume there are n risky assets with random rates of return $r_1, r_2, \ldots, r_n$ among which the investor can allocate his wealth. The investor can also invest in a riskless asset offering a sure rate of return s_0. Denote by x_i the amount invested in i-th risky asset and $(w_0 - x_1 - x_2 - \ldots - x_n)$ the amount invested in the riskless asset. Then the final wealth of the investor in the next period becomes

$$w_1 = s_0\left(w_0 - \sum_{i=1}^{n} x_i\right) + \sum_{i=1}^{n} r_i x_i$$

$$= s_0 w_0 + \Sigma m_i x_i, \qquad m_i = r_i - s_0.$$

The objective is to maximize over the decision vector $x = x_i$ the expected utility function $E[U(w_j)]$, where $U(\cdot)$ is the utility function of the investor. For convenience no other conditions are imposed on the decision vector x, which implies the assumption that the joint probability distribution of the random return vector r and the minimizing values of x are such that the values of w_1 belong to the domain of the utility function $U(\cdot)$. In the MV theory the expected utility function comprises the variance term $x'Vx$, where V is the variance covariance matrix for $(r - s_0)$. In the intertemporal case with a fixed horizon $t = 0, 1, \ldots, T$ the dynamic system equation becomes

$$w_t = s_t w_{t-1} + m_t' x_t + \varepsilon_t, \qquad \varepsilon_t \sim N(0, \sigma_\varepsilon^2) \tag{7.2.9}$$

with w_0 given. The objective may be either to maximize the expected utility $\mathrm{E}[U(w_t)]$ of the terminal wealth or, to minimize the sum total of a quadratic loss function. Taking the latter case we may write

$$J = \min \mathrm{E}\left[\theta \exp\left\{\frac{\theta}{2}\sum_{t=0}^{T-1} x_t' V_t x_t\right\}\right] \tag{7.2.10}$$

where the constant term $\theta/2$ specifies the De Finetti-Arrow-Pratt measure of absolute risk aversion for $\theta > 0$. Thus the higher the positive value of θ, the greater is the degree of absolute risk aversion i.e., the different risk classes are then characterized by different values of θ from low to medium and from medium to high. Note that in keeping with the MV theory only the variance term has been considered in the objective function (7.2.10), where V_t may be interpreted as inclusive of a discount factor i.e., $V_t = \rho_t \tilde{V}_t$ for a constant discount rate ρ. Two other assumptions are made as follows: one is that the stochastic component ε_t is normally and independently distributed with mean zero and constant variance σ_ε^2. The second is that a conditional mean estimate $w_t = \mathrm{E}[w_t | I_{t-1}]$ of w_t can always be made from the observed data, given the information set I_{t-1} containing all information up to t-1. It is well known that if all noises are assumed Gaussian, then w_t is nothing but the optimal Kalman filter of optimal control theory.

On applying the optimal control theory methods to (7.2.9) and (7.2.10), developed in some detail by Whittle (1990) we may easily derive the optimal linear decision rules as:

$$x_t^* = -C_t \hat{w}_t, \qquad \hat{w}_t = \mathrm{E}[w_t | I_{t-1}] \tag{7.2.11}$$

where for $t = 0, 1, \ldots, T-1$

$$C_t = [V_t + m_t \tilde{h}_{t+1} m_t']^{-1} m_t \tilde{h}_{t+1} s_t$$

$$\tilde{h}_t = h_{t+1}[1 + \theta\sigma_\varepsilon^2(1 - \theta h_{t+1})^{-1}]$$

$$h_t = s_t^2 \tilde{h}_{t+1}(1 - k_t \tilde{h}_{t+1})$$

$$k_t = m_t'(V_t + m_t \tilde{h}_{t+1} m_t')m_t$$

Also we have

$$J(w_t) = f_t \exp[w_t' h_t w_t / 2]$$

$$x_t^{*\prime} V_t x_t^* = C_t' V_t C_t \hat{w}_t^2$$

$$f_t = f_{t+1} \sigma_\varepsilon (|1 - h_{t+1}|)^{1/2}, \qquad f_T = 1.$$

In the steady state one obtains

$$x^* = -(V - m\tilde{h}\, m')^{-1} m \tilde{h}\, s \hat{w}$$

where $\tilde{h}$ is negative if $\sigma_\varepsilon^2 + s^2 < 1$.

Two general cases of this dynamic formula are of special importance here. One arises when the error term ε_t in (7.2.9) is heteroscedastic i.e., it has a non diagonal variance covariance matrix V_ε say. Then the equations for updating $\tilde{h}_t$ and f_t are only changed as follows

$$\tilde{h}_{t+1} = h_{t+1} + \theta h_{t+1} V_\varepsilon (V_\varepsilon^{-1} - \theta h_{t+1})^{-1} h_{t+1}$$

$$f_t = f_{t+1} |V_t|^{-1} [|(V_\varepsilon^{-1} - \theta h_{t+1})^{-1}|]^{1/2}, \qquad f_T = 1.$$

The second case arises when different asset classes are first considered e.g., different classes of mutual funds such as growth funds, balanced funds and income funds and then within each class the investor decides on an optimal allocation vector x. In this case the term w_t becomes a vector but the formal structure of the solutions (7.2.11) remains the same. This framework is most suitable when there is switching of investments from one group, of mutual funds to another.

Several implications of the optimal decision rule (7.2.11) may now be discussed. First of all, it is more general than the MV decision rule. It is adaptive in two ways. One is that it is time varying and hence updating through new information contained in the information set I_t occurs sequentially over time. Secondly, it is risk sensitive since the changing variance parameters V_t and V_ε and the risk aversion index θ influence the optimal decision rule at each time point t. Since a negative value of θ implies a risk seeking rather than risk averting attitude, one could empirically let the observed data determine its sign. Secondly, the investor's adjustment behavior over time may be easily captured in this formulation by rewriting the objective function (7.2.10) as:

$$J = \min \mathrm{E}\left[\theta \exp\left\{\frac{\theta}{2}\sum_{t=0}^{T-1}\left[(w_t - a_t)^2 + x_t'V_t x_t\right]\right\}\right] \qquad (7.2.12)$$

where a_t may represent the target level of wealth. In this case the only change in the optimal rule (7.2.11) is the definition of h_t in the feedback multiplier C_t as follows:

$$h_t = 1 + s_t^2 \tilde{h}_{t+1}(1 - k_t \tilde{h}_{t+1})$$

Since the linear part of the exponential objective function yields a quadratic loss function, it is clear that the MV efficiency frontier would be a special case of this exponentially quadratic objective function. Finally, consider the divergence between the observed state variable w_t and its optimal values w_t^* where

$$w_t = w_t^* - u_t, \qquad u_t \geq 0$$

$$= -\bar{u}_t + s_t w_{t-1} + m_t' x_t + \varepsilon_t, \qquad \bar{u}_t = \mathrm{E}[u_t].$$

In this case the steady state values of variance $\sigma^2 = x'Vx$ can be written as:

$$\sigma^2 = C'VC(1 - s)^2[c^2 - 2\bar{u}c + \bar{u}^2], \quad \bar{c} = m'x^* \qquad (7.2.13)$$

Clearly $\partial\sigma^2/\partial c$ is negative (positive) if $\bar{u}$ is greater (less) than the expected return $c = m'x^*$. Thus it is possible due to asymmetry in the risk sensitive behavior that the mean and variance may at times be negatively correlated. In other words the risk reaction in a pessimistic market may be quite different from that in an optimistic market. From an empirical standpoint then one could test for asymmetry through the estimation of the variance frontier equation (7.2.13).

7.2.3 Empirical applications.

For empirical applications of the adjustment cost models in mean variance efficiency we consider three data sets. One is provided by the time series monthly data for five market returns taken from the CRSP tape for the period: January 1965 through December 1975. This period was selected for two reasons. One is that the same data set were used in previous studies for analyzing the risk return relationships for three types of mutual funds in order to test the significance of the third moment of the return distribution. Secondly, the above period includes both bearish (January 1973 through December 1974) and bullish subperiods (January 1975 through December 1975), so that one could test the variations if any of the impact of the third moment on the conditional variance in these two periods. The second data set comprises monthly returns data for five mutual funds over the bearish subperiod January 1973 through December 1974. These five funds are selected one each from Wiesenberger's classification into maximum capital gain fund (A), growth fund (G), mixed fund with growth and income objectives (M), income fund (I) and balanced fund (B). The selected funds are: Admiralty Fund-Growth series (A_1), All State Enterprises Stock Fund (G_1), Bullock Fund (M_1), Admiralty Fund-income series (I_1) and Axe-Houghton Fund-Balanced Series (B_1). The third data set is for the monthly return series for the market indices i.e. VW (Value Weighted index), and the Standard and Poor 500 (S&P 500) for the overall period October 1976 through December 1991 divided into two sample periods. Sample I (October 1976 through July 1982) and Sample II (August 1982 through December 1991). According to Wissemberger reports, the first sample period is a bearish market while the second sample a bullish market. Since a bullish market is more likely to exhibit increased volatility, the results from the second sample of the third data set are presented in Tables 5 to 7, although the bearish market is considered as a reference for comparison.

For the first data set, the five market indices used are as follows: (a) Equally weighted market return including New York Stock Exchange (NYSE) and American Stock Exchange (AMEX) securities, (b) value weighted market return for NYSE and AMEX securities and (c) Standard and Poor 500 (S&P 500) index. The rates of return for these five indices include the dividends paid per share and are computed in the usual way by taking the natural logarithm of one plus the monthly return. The above five indices are fairly comprehensive in coverage of the overall market behavior. A direct test of the risk return relationship for the market indices may be performed by a linear regression of the standard deviation $\sigma_t(c)$ of return on the observed mean return c_{t-1} in the last period. Note that in the empirical regressions the lagged variable c_{t-1} is used as a predetermined variable as is done in Arch models. Three types of sampling framework are considered, e.g., (a) fixed sample size, (b) moving average samples and (c) increasing sample sizes. For the fixed sample size the entire 10 year period is divided into 30 four month subperiods, so that the first subperiod is January 1965 through April 1965, the second from May 1965 through August 1965 and so on. For the moving average samples the means and standard deviations are computed for the 4-month moving average samples, so that the first sample runs from January 1965 through April 1965, the second sample from February 1965 through May 1965 and so forth. Finally, the increasing sample size test is based on the means and standard deviations computed by gradually increasing the sample sizes by 4 months beginning with January 1965. Thus the first subperiod is from January 1965 through April 1965, the second subperiod from January 1965 through August 1965 and so forth. The linear regression estimates of the standard deviation frontier based on the COLS method are given in Table 1. Clearly in all the cases the standard deviation is negatively related to the mean implying a perverse mean variance relationship. Of the three sampling frameworks the case of increasing samples produces in a statistical sense the best results in the sense of goodness of fit and significance of the estimated coefficients, i.e., the slope coefficients are significant at 1% level of the t-test. Also this has more intuitive appeal since it tests the persistence of mean variance relationship when sample size increases.

One possible reason for the negative relationship between mean and variance is that higher order moments such as skewness and kurtosis are ignored here. In studies analyzing British mutual funds (i.e., unit trusts) Saunders, Ward and Woodward (1980) have found that the more risky the trust the lower are its returns on the average. Also the investors in U.K. appear to be less risk averse, since the trading volume in these funds did not decline despite the poor performance of the high-risk trusts.

The most interesting implication of the estimates of Table 1 is that it lends support to the hypothesis of variance declining monotonically with the mean, although the return distribution is not restricted to be normal alone. So long as the return distribution belongs to the class of log concave densities and the investors are seeking higher and higher levels of average returns, one could expect a negative relationship between the mean and variance. As we have seen that the adjustment cost model predicts this type of behavior.

For some additional tests on the negative relationship between the conditional variance and the mean, a more recent sample period August 1982 through 1991 was chosen. This period has been characterized as a bullish market by the Weisenberger Reports. In this market the active investors tend to dominate so that the return distribution belongs to the log-concave class in a conditional sense. The results of the linear regression test based on the COLS method are as follows:

Market Index	Intercept	c_{t-1}	c_{t-1}^2	c_{t-1}^3	$\bar{R}^2$	DW
Equally weighted return (NYSE)	0.0012**	-0.0385*	3.236**	-43.9**	0.53	1.16
Value weighted return (NYSE)	0.0015**	-0.0628**	3.765**	-59.3**	0.69	1.25
S&P500	0.0015**	-0.0441**	3.164**	-58.1**	0.56	1.17

Note: One and five percent levels of t are marked by one or two asterisks respectively.

Clearly the negative mean variance relationship is persistently observed for this bullish market period. To test further the nonlinear relationships between the variance and the mean, Tables 2 and 3 have reproduced the empirical estimates of the variance frontier model (7.2.5), where due to multicolinearity problems regressor variables have been retained up to the cubic terms. Note that the regression coefficient γ_3 for the cubic term c_{t-1}^3 is consistently negative for all the five market returns and they are highly significant at 1% level. This pattern is not altered when the logarithm of variance is used as a dependent variable. Also in terms of the standard measure of skewness the equally weighted returns are found to be more skewed than the value-weighted return for the whole period. Hence this supports the findings by Beedles (1979) that the measured asymmetry of the equally weighted index is larger than the value weighted index.

Another interesting finding is that variances are monotonically decreasing functions of c for the value weighted indexes when data is removed between January 1973 and December 1974, considered as a bearish market period. On the contrary, variances calculated from equal weighted return are not strictly monotonically decreasing in c during the same period although they show approximately a monotonically negative relationship. Test results differ whether one includes or excludes this bearish market period. Finally, the estimates of the cubic variance function σ_t^2 show that it is not uniformly convex in c, since the coefficient γ_3 determines whether it is convex or concave. This suggests the need for generalizing the concept of risk aversion by allowing for adjustment cost. However it is clear from the estimated coefficient γ_3 of the cubic term that the impact of skewness on variance is highly significant.

Next one may consider the second data set for five major mutual funds and estimate variance function models which may be considered as the steady state versions of adjustment cost transformation of the MV efficiency models. The estimates which are presented in Table 4 clearly show two broad results. The first is that the conditional variance σ_t^2 is negatively correlated with the mean levels of return c_{t-1} for each mutual fund and this relationship is highly significant statistically. Moreover the regression coefficient of c_{t-1} for growth fund (Gl:-0.022) is higher in absolute value than the income and balanced funds. Secondly, the variance function is strictly convex in the mean return level i.e.,

$$\sigma_t^2 = a_0 - a_1 c_{t-1} + a_2 c_{t-1}^2, \qquad a_0, a_1, a_2 > 0$$

from which it follows that there is a level c of c_{t-1} at which σ_t^2 can be minimized i.e.,

$$c = a_1(2a_2)^{-1}.$$

Since the intercept term a_0 is positive and statistically significant for all cases, it is clear from the variance function formulation (7.2.5) that the adjustment cost function may be estimated as

$$\sigma_A^2 = \bar{u} - \hat{a}_0, \qquad \bar{u} > \hat{a}_0.$$

A second way to analyze the mutual fund data set is to apply the two-step estimation procedure adopted in the ARCH model discussed before. The return series $\{y_t\}$ here is decomposed into conditional mean and conditional variance i.e.,

$$y_t = E_{t-1}[y_t] + e_t; \qquad \sigma_t^2 = E_{t-1}[\sigma_t^2]$$

and then one computes the linear regression of the estimated variances $\hat{\sigma}_t^2$ on $\hat{e}_{t-1}^2$ and $\hat{\sigma}_{t-1}^2$. The following results then emerge:

A_1: $\hat{\sigma}_t^2 = 0.0001 - 0.002\,\hat{e}_{t-1}^2 + 0.962^{**}\,\hat{\sigma}_{t-1}^2$; $\bar{R}^2 = 0.92$

G_1: $\hat{\sigma}_t^2 = 0.0002^{*} - 0.011\,\hat{e}_{t-1}^2 + 0\,939^{**}\,\hat{\sigma}_{t-1}^2$; $\bar{R}^2 = 0.92$

M_1: $\hat{\sigma}_t^2 = 0.0 + 0.002\,\hat{e}_{t-1}^2 + 0.956^{**}\,\hat{\sigma}_{t-1}^2$; $\bar{R}^2 = 0.92$

I_1: $\hat{\sigma}_t^2 = 0.00 + 0.012\,\hat{e}_{t-1}^2 + 0.941^{**}\,\hat{\sigma}_{t-1}^2$; $\bar{R}^2 = 0.93$

B_1: $\hat{\sigma}_t^2 = 0.0002^{*} - 0.005\,\hat{e}_{t-1}^2 + 0.769^{**}\,\hat{\sigma}_{t-1}^2$; $\bar{R}^2 = 0.59$

Clearly the coefficient of $\hat{\sigma}_{t-1}^2$ is highly significant and very close to unity. This implies that market volatility measured by the conditional variance is persistent or nearly so. On applying this method to the five market indices analyzed before one obtains a similar result i.e., that the market volatility is persistent and statistically significant.

Next we consider the third data set for Sample II covering the bullish period: August 1982 through December 1991. For this sample set of returns, the same method is followed for regressing the conditional variances $\sigma_t^2 = E_{t-1}[e_t^2]$ on the conditional means $c_t = E_{t-1}[y_t]$, where the observed market return is $y_t = E_{t-1}[y_t] + e_t$. To test the persistence of the variance process we consider the second order linear model

$$\sigma_t^2 = a_0 + a_1\sigma_{t-1}^2 + a_2\sigma_{t-2}^2. \tag{7.2.14}$$

Likewise for the skewness process $s_t = E_{t-1}[e_t^3]$:

$$s_t = d_0 + d_1 s_{t-1} + d_2 s_{t-2}. \tag{7.2.15}$$

If the variance process measuring market volatility is persistent, then one would expect the sum $(a_1 + a_2)$ to equal or exceed one. Likewise for the skewness process persistence would require the sum $(d_1 + d_2)$ to equal or exceed one.

Two striking features of Table 5 are the strong significance of the coefficient of the cubic term c_{t-3}^3 and the negative coefficient associated with the mean c_{t-1}. The estimated coefficients alternate in sign starting with the negative coefficient. This implies that both mean and skewness have strong negative impact on the conditional variance. This asymmetry effect presents a strong contrast to the standard mean variance hypothesis of the capital asset pricing model, where increasing mean is associated with higher risk. Secondly, it is easy to check that the variance is a declining function of the mean, if the following condition $c_{t-1} > 0.67\ b_2/b_3$ holds, where b_2 and b_3 are the coefficients of c_{t-1}^2 and c_{t-1}^3. Since this holds here, one would expect a negative correlation between mean and variance. However if the above condition does not hold, then the variance may exhibit the standard feature, i.e., an increasing function of the mean. The following results for sample I (bearish market) tend to show that the bearish market period fails to exhibit the property of variance declining with the mean.

Return	b_0	b_1	b_2	b_3	$\bar{R}^2$
VW	0.0017**	0.0246	0.0686	-25.13	0.05
EW	0.0417**	-0.0443	-2.5071	45.80	0.06
S&P	0.0016**	0.0229	-0.1761	-22.40	0.04

The estimates of the variance persistence model (7.2.14) shows that the first order coefficient $\hat{a}_1$ is close to one for VW and S&P500, but the second coefficient $\hat{a}_2$ is uniformly negative, though not statistically significant at the 5% level. Similar estimates for the skewness persistence model (7.2.15) show that there is no empirical evidence of skewness persistence in the bullish period.

Table 1

Linear Standard Deviation Frontier (COLS Method)

Market Index	Intercept (a)	Intercept (b)	Intercept (c)	Slope (a)	Slope (b)	Slope (c)	$\bar{R}^2$ (a)	$\bar{R}^2$ (b)	$\bar{R}^2$ (c)
Equally weighted return;									
NYSE	0.051**	0.050**	0.061**	-0.231	-0.142*	-1.067**	.08	.04	.80
NYSE+AMEX	0.057**	0.056**	0.072**	-0.190	-0.104	-0.981	.07	.02	.75
Value weighted return;									
NYSE	0.039**	0.039**	0.042**	-0.305	-0.290*	-1.022**	.09	.10	.40
NYSE+AMEX	0.41	0.039**	0.043*	-0.314*	-0.288*	-1.061**	.10	.11	.43
S&P 500 Index;	0.037**	0.037*	0.038**	-0.327	-0.335*	-0.827**	.09	.13	.31

(a) = Fixed samples, (b) = Moving average, (c) = Increasing samples

Notes:

1. One and two asterisks denote significant t-value at 5 and 1 percent calculated from the estimated asymptotic standard errors
2. Adjusted R^2 values are denoted $\bar{R}^2$

Table 2

Parameter Estimates for the Stochastic Variance Frontier Models (COLS)

Market Index	Intercept	c_{t-1}	c_{t-1}^2	c_{t-1}^3	$\overline{R}^2$	DW
Equally weighted return;						
NYSE only	0.004**	-0.023**	-	-	0.85	2.29
	0.003**	-0.050**	0.248**	-	0.94	2.03
	0.004**	-0.077**	0.828**	-3.311**	~.95	2.41
NYSE+AMEX	0.004**	-0.025**	-	-	0.83	1.69
	0.004**	-0.054**	0.221**	-	0.94	2.07
	0.005**	-0.086**	0.797**	-2.750**	0.96	2.31
Value weighted return;						
NYSE only	0.002**	-0.004*	-	-	0.09	1.91
	0.003**	-0.041**	0.383**	-	0.90	2.07
	0.003**	-0.063**	0.925**	-3.442**	0.96	2.69
NYSE+AMEX	0.001**	-0.006**	-	-	0.16	1.79
	0.002**	-0.040**	0.357**	-	0.87	2.01
	0.002**	-0.066**	1.000**	-4.056**	0.96	2.42
S&P 500 Index;						
	0.001**	-0.002	-	-	0.01	1.95
	0.002**	-0.036**	0.384**	-	0.86	2.03
	0.002**	-0.057**	0.961**	-3.938**	0.95	2.29

Notes:

1. One and two asterisks denote significant t-value at 5 and 1 percent calculated from the estimated asymptotic standard errors
2. Adjusted R^2 values are denoted $\overline{R}^2$

Table 3

Parameter Estimates for the Stochastic Variance Frontier Models (COLS Method; Log Variance)

Market Index	Intercept	c_{t-1}	c_{t-1}^2	c_{t-1}^3	$\overline{R}^2$	DW
Equally weighted return;						
NYSE only						
- log variance	-5.02**	-33.04**	244.89**	-1590.51**	0.96	2.34
NYSE+AMEX						
- log variance	-5.01**	-31.58**	261.19**	-1369.81**	0.96	2.42
Value weighted return;						
NYSE only						
- log variance	-6.14**	-59.10**	832.92**	-2682.92**	0.95	2.51
NYSE plus AMEX						
- log variance	-6.10**	-60.65**	855.78**	-3094.30**	0.95	2.19
S&P 500 Index;						
- log variance	-6.26**	-58.03**	973.90**	-3998.41**	0.95	2.06

Notes:

1. Two asterisks denote significant t-value at 1 percent calculated from the asymptotic standard errors of the estimates.
2. Adjusted R^2 values are denoted $\overline{R}^2$

Table 4

Variance Function Models for Mutual Funds (COLS Method)

Type of Fund	Intercept	c_{t-1}	c_{t-1}^2	$\bar{R}^2$	F-value
A_1: σ_t^2	0.002**	-0.014**	-	0.71	28.6
	0.002**	-0.031**	0.271**	0.95	111.8
G_1: σ_t^2	0.004**	-0.022**	-	0.90	104.9
	0.004**	-0.032**	0.073	0.91	58.9
M_1: σ_t^2	0.002**	-0.009**	0	0.54	15.0
	0.003**	-0.033**	0.231**	0.94	102.1
I_1: σ_t^2	0.001**	-0.005*	-	0.15	3.1
	0.001**	-0.035**	0.536**	0.91	64.3
B_1: σ_t^2	0.001**	-0.012**	-	0.97	444.3
	0.001**	-0.017**	0.091**	0.99	628.9

Notes:

1. One and two asterisks denote significant t-value at 5 and 1 percent levels
2. Adjusted R^2 values are denoted $\bar{R}^2$
3. F-values denote F statistics for the estimated errors.

Table 5

Variance Frontier Estimates
(Sample II: August 1982 - December 1991)

Return	Intercept	c_{t-1}	c_{t-1}^2	c_{t-1}^3	$\bar{R}^2$	DW
VW	0.0015**	-0.0628**	3.7647**	-59.34**	0.69	1.25
EW	0.0012**	-0.0385*	3.2361**	-43.99**	0.53	1.16
S&P	0.0015**	-0.0441**	3.1640**	-58.12**	0.56	1.17

Notes:

1. One and two asterisks denote significant t-value at 5 and 1 percent levels
2. Adjusted R^2 values are denoted $\bar{R}^2$
3. DW denotes the Durbin-Watson statistics.

Table 6

Variance Persistence Model
(Sample II: August 1982 - December 1991)

Return	a_0	a_1	a_2	$\overline{R}^2$	DW
VW	-0.0004**	0.986**	-0.184	0.74	2.04
EW	-0.0005	0.907**	-0.124	0.65	2.02
S&P	-0.0004*	1.009**	-0.204	0.71	2.05

Notes:
1. One and two asterisks denote significant t-value at 5 and 1 percent levels
2. Adjusted R^2 values are denoted $\overline{R}^2$
3. DW denotes the Durbin-Watson statistics.

Table 7

Skewness Persistence Model
(Sample II: August 1982 - December 1991)

Return	d_0	d_1	d_2	$\overline{R}^2$	DW
VW	-0.007	0.7261**	0.002	0.54	2.02
EW	-0.035	0.6011**	0.172	0.54	2.01
S&P	-0.012	0.8204**	-0.098	0.57	2.03

Notes:
1. One and two asterisks denote significant t-value at 5 and 1 percent levels
2. Adjusted R^2 values are denoted $\overline{R}^2$
3. DW denotes the Durbin-Watson statistics.

7.3 Econometric estimation of controlled systems

A deterministic dynamic system where the parameters are all known and there are no errors of disturbance may be viewed in two ways in a control framework, either as a feedback system, or as an optimal system. Assuming a linear discrete time dynamic system with constant parameters, the feedback system may be written as:

$$x(t) = Ax(t-1) + Bu(t) \tag{7.3.1}$$

where $x(t) \in \Re^{m}$ and $u(t) \in \Re^{n}$ are the state and control vectors respectively, and A, B are constant parameters assumed to be known. The system is subject to the initial conditions $x(0) = x_0$ (given), with an optimizing objective function. For example consider a simple case of a linear quadratic problem with the following objective function:

$$\min J = \tfrac{1}{2}\sum_{t=1}^{T}\left[x'(t)Qx(t)+u'(t)Ru(t)\right] \tag{7.3.2}$$

where for simplicity we assume that the matrix R is positive definite, Q is symmetric semidefinite and the state control variables are in deviation terms i.e., $x(t) = \mathrm{x}(t) - \mathrm{x}^{o}(t)$, and $u(t) = \mathrm{u}(t) - \mathrm{u}^{o}(t)$, where $\mathrm{x}^{o}(t)$ and $\mathrm{u}^{o}(t)$ are the desired values of the state and control variables, known or preassigned by the decision maker. As we have seen in previous chapters, the optimal control rule can easily be computed here as a feedback rule:

$$u^{*}(t) = K(t)x(t-1) + k(t) \tag{7.3.3}$$

where $K(t)$ and $k(t)$ can be recursively computed from the Riccati equations as follows:

$$K(t) = -[R + B'(Q + P(t+1))B]^{-1}B'(Q + P(t+1))A$$

$$k(t) = -[R + B'(Q + P(t+1))B]^{-1}[B'(Q\mathrm{x}^{o}(t) + h(t+1)) - R\mathrm{u}^{o}(t)]$$

$$h(t) = [A + BK(t)]'[Q + P(t+1)\mathrm{x}^{o}(t) - h(t+1)] - K'(t)RK(t).$$

In the infinite horizon case, if the dynamic system is controllable, then $P(t)$ becomes a constant and hence the gain matrix $K(t)$ becomes a constant. Furthermore, if the desired levels $\mathrm{x}^{o}(t)$ and $\mathrm{u}^{o}(t)$ are fixed, then the tracking gain vector $k(t)$ also becomes a constant. In this case, the linear optimal control rule takes the simple form:

$$u^{*}(t) = Kx(t-1) + k \tag{7.3.4}$$

On substituting this in the state equation (7.3.1), one obtains the optimally controlled system:

$$x^*(t) = (A + BK)x^*(t\text{-}1) + Bk \tag{7.3.3}$$

where the asterix denotes the optimal values.

The control problems become stochastic whenever any of the following situations are present, either singly or jointly e.g.,

a/ The system parameters A, B, Q, R are unknown, but the sample data are available on $x(t)$ and $u(t)$ to estimate them.

b/ There are disturbances in the state dynamics e.g., the LQG model assumes $x(t) = Ax(t\text{-}1) + Bu(t) + \varepsilon(t)$, where $\varepsilon(t)$ is a white noise Gaussian process, independently distributed from $u(t)$.

c/ The initial and terminal conditions $x(0)$ and $x(T)$, or the desired levels $x^o(t)$ and $u^o(t)$ are not known with certainty.

The most common approach in economic modeling is to estimate the parameters of the state dynamics (7.3.6) by maximum likelihood or least squares, and then compute the optimal control rules (7.3.3) and (7.3.4) after assuming that the parameter estimates are known quantities. The two stage separation of the stochastic control problem is appropriate only when the variance of the system error $\varepsilon(t)$ is very low, and the parameters A, B are nearly constant over the planning horizon. Otherwise, there may arise time inconsistency problems as discussed in chapter 5.

A second approach employs an adaptive estimator developed by Astrom and Wittenmark (1973), Ljung (1987) and others, which updates the parameter estimates during the control period. Recently, Ljung (1987) developed a recursive least squares algorithm, where an adaptive discounting factor is incorporated. Since the discounting factor may be interpreted as weights inversely proportional to the variance of equational errors I the state dynamics, this method may be termed weighted recursive least squares (WRLS) algorithm. This method can efficiently handle the situation when parameters are changing more or less rapidly, in response to new observations.

A third approach employs the method of joint estimation and control. Two variants of the method are frequently used in applied control literature. One is the direct method proposed by Ljung et. al. (1974), and the other is an indirect method proposed by Chow (1975). In the direct method of Ljung et. al., one writes the control law as a linear structure:

$$u(t) = Fx(t\text{-}1) + Ge(t) \tag{7.3.6}$$

where $e(t)$ is an error term and the control rule is not necessary optimal. This control structure together with the system (7.3.1) are then incorporated in a simultaneous equation structure:

$$u(t) = Fx(t\text{-}1) + Ge(t)$$

$$x(t) = M(t) + Ne(t) \tag{7.3.7}$$

where $M = [\mathrm{I} - A\mathrm{L}]^{-1}B$, $N = [\mathrm{I} - A\mathrm{L}]^{-1}$ and L is a backward operator, i.e. $\mathrm{L}x(t) = x(t\text{-}1)$. If the joint distribution of $e(t)$ and $\varepsilon(t)$ is multivariate normal with fixed means and variances, then the method of joint maximum likelihood may be applied to obtain the estimates of the parameters. When the errors $e(t)$ and $\varepsilon(t)$ are not independent, the method of seemingly unrelated regression can be utilized. For non normal distributions, nonlinear maximum likelihood procedures have to be followed. Two points have to be noted in this approach. First, the separation of the two steps, estimation in the first stage, and then applying control in the second stage is not followed here. Second, it shows the contrast between optimal control and optimal estimation in a striking fashion. Thus the first equation system of (7.3.7) may be interpreted as a feedback control, which is approximately optimal as in a LQG model. The second equation system however is in a typical reduced form, where the endogenous variables $x(t)$ are exogenous functions of the exogenous variables $u(t)$.

The indirect method proposed by Chow (1975) recognizes from the outset that some modification of the quadratic objective function (7.3.2) is needed to take account of parameter uncertainty. The modified objective function takes the following form:

$$\min J = \mathrm{E}\left[\sum_{t=1}^{T}\left[\tfrac{1}{2}x'(t)W(t)x(t) + h'(t)x(t)\right]\right] \tag{7.3.2}$$

where $x(\mathrm{t}) = \mathrm{x}(t) \text{-} \mathrm{x}^{\mathrm{o}}(t)$ as before, and the elements of the weight matrix $W(t)$ represent cost weighting placed on the correlation between past weights on outputs and the elements of vector $h(t)$ denote a set of linear weights on the output deviations. The crucial assumption is that the weights are known or specified. But this procedure is highly nonlinear and some numerical procedures for approximation are necessary. The experience of applying this method in Keynesian macroeconomic models by Chow is that the computational burden is very heavy and perhaps not worth following unless the parameter uncertainties are very large.

One may mention another method, which assumes that the rational expectations hypothesis holds so that the observed data conform to the optimal feedback rule (7.3.4) implicitly. Then the controlled system model (7.3.5) can be estimated by the method of instrument variables, provided the distribution of the error structure $\xi(t)$ can be specified e.g.,

$$x^*(t) = Cx^*(t\text{-}1) + c + \xi(t). \tag{7.3.8}$$

In case the errors are normally distributed with a zero mean and a constant variance, the method of weighted least squares may again be applied. Some empirical applications by Sengupta and Sfeir (1979) to the aggregate inventory and output data of several industries in the US show that this type of model performs very well in a forecasting sense. Note however that one cannot recover the estimates of the original structural parameters A and B for the estimates of the parameter matrix C, since this model is not identified. The forecasting performance of such models (7.3.8) is comparable to the recent VAR (Vector AutoRegressive) models applied in dynamic models.

Finally, we have to note that the initial $x(0)$ and terminal $x(T)$ conditions have to be estimated as they are assumed unknown. The starting point is usually known, but the terminal point is often unknown. One way to handle the terminal point is to adjoin it as a separate constraint and then maximize the relevant log-likelihood function subject to this constraint. Alternatively, one can adjoin a penalty function to the original objective function e.g., in the case of (7.3.2), we adjoin the term $\frac{1}{2}[x(t) - d]'[x(t) - d]$, where $x(T) = d$ is the constraint and $x(T)$ now has to be estimated. This aspect is discussed in some detail in the next section.

We now consider two specific examples of econometric estimation of a system. One is the WRLS algorithm, which is often applied to international stock market models where the parameters frequently change over time. The second illustrates the interaction between estimation and control, which is sometimes applied in modern control theory as an optimal design problem.

7.3.1 WRLS Algorithm and its Implications.

Consider the vector process y_t of dimension p admitting a linear representation:

$$y_t = \theta_t' z_t + v_t \tag{7.3.9}$$

where y_t is second order stationary with time varying parameter matrix θ_t and an m-dimensional vector of regressors z_t, which may contain the lagged values of output y_t and also the control variables. The vector v_t is assumed to have components identically and independently distributed with mean zero and variance σ_v^2. Assuming for simplicity that $p = 1$, Ljung (1987) showed that the solution to the weighted least square criterion

$$\hat{\theta}^2 = \underset{\theta_t}{\arg\min}\left[\sum_{k=1}^{t} \beta(t,k)\{y(k) - z'(k)\theta_t\}^2\right]$$

is given by $\hat{\theta}_t = R^{-1}(t)g(t)$ where $R(t) = \sum_{k=1}^{t} \beta(t,k)z(k)z'(k)$ and $g(t) = \sum_{k=1}^{t} \beta(t,k)z(k)y(k)$. Now the weighting scheme $\beta(t,k)$ can be chosen in different forms. Like the Box-Cox transformation, the scheme can be defined as an increasing or decreasing sequence. For example, suppose the weights satisfy the following property $\beta(t,k) = \lambda(t)\beta(t\text{-}1,k)$, $1 \leq k \leq t\text{-}1$, and $\beta(t,t) = 1$. Then one can write: $R(t) = \lambda(t)R(t\text{-}1) - z(t)z'(t)$, $g(t) = \lambda(t)g(t\text{-}1) - z(t)y(t)$. Hence $\hat{\theta}_t = R^{-1}(t)g(t)$ which can also be expressed as $\hat{\theta}_t = \hat{\theta}_{t-1} + R^{-1}(t)z(t)[y(t) - z'(t)\hat{\theta}_{t-1}]$. Then the RLS algorithm can be iteratively applied as:

$$\hat{\theta}_t = \hat{\theta}_{t-1} + R^{-1}(t)z(t)[y(t) - z'(t)\hat{\theta}_{t-1}]$$

$$R(t) = \lambda(t)R(t\text{-}1) + z(t)z'(t) \tag{7.3.10}$$

over time $(t = 1, 2, \ldots, \mathrm{T})$. Note that $\lambda(t)$ can also be selected as a constant, or a sequence of constants, and an optimal subset may be chosen as that subset that maximizes the overall forecasting performance, measured by the squared multiple correlation coefficient. Two implications of this algorithm are most important in economic applications. One is that the parameters are allowed to change over time through a discounting factor, thus providing a convenient means for recognizing past information to a specific degree. Also the results of Bera et. al. (1992) show that the VRLS algorithm is capable of reducing the conditional heteroscedasticity, which is frequently used in models of stock market volatility e.g., ARCH and GARCH models. Secondly, one has to note that the basic similarity of this recursive estimate $\hat{\theta}_t$ in (7.3.10) with that obtained with the Kalman filter analyzed in previous chapters. Thus one can use this approach in estimating the time point when the parameters change from one level to another e.g., structural change due to technological innovation. In order to illustrate the detection test for parameter change, consider an autoregressive scalar process AR(1);

$$y(t) = M(t)\theta(t) + v(t) \tag{7.3.11a}$$

where $M(t) = [y(t\text{-}1)]$, $\theta'(t) = [\theta_1(t)]$, $\mathrm{E}[v(t)] = 0$, $\mathrm{E}[v(t)v(t\text{-}k)] = R$ if $k = 0$, and zero otherwise, and the sequence of random variables $v(t)$ is normally and independently distributed. On using the Kalman filter estimate $\hat{\theta}(t|t\text{-}1)$, the filter's residual is

$$e(t|t\text{-}1) = y(t) - M(t)\hat{\theta}(t|t\text{-}1) = y(t) - \hat{\theta}_1(t|t\text{-}1)y(t\text{-}1),$$

where $\hat{\theta}(t|t\text{-}1)$ is the filter's estimate for the conditional expected value of $\theta(t)$, given that there is available data up to time t - 1. Suppose the parameter $\theta_1(t)$ has a step change at time $t = \tau + 1$ i.e.

$$\theta_1(t) = \theta_o(t) \quad \text{when} \quad t \leq \tau, \qquad \theta_1(t) = \theta_n(t) \quad \text{when} \quad t > \tau \tag{7.3.11b}$$

where the subscripts o and n stand for old and new values respectively. Suppose the current time is $t < \tau$ i.e. change has not yet occurred and assume that the parameter estimate is computed with the data prior to change. Then by using (7.3.11b) the filter's residual is given by:

$$e(t|t\text{-}1) = y(t) - \hat{\theta}_o(t|t\text{-}1)y(t\text{-}1), \qquad t < \tau, \qquad (7.3.11c)$$

where $\hat{\theta}_o(t|t\text{-}1)$ is the Kalman filter estimate of $\theta_o(t)$. Suppose now that a parameter change has actually occurred, but the forecaster is still computing prediction from old estimates. Then the estimates so obtained are completely biased. To detect this situation we note that if a postulated change has occurred, then the current time is $t > \tau$ and consequently $y(t)$ can be expressed in terms of the new parameter. However since $y(t)$ can also be expressed as a linear combination of previous values, then it can be expressed as a mixture of old and new parameters. Thus the error equation (7.3.11c) can be expressed as:

$$e(t|t\text{-}1) = \theta_n(t)y(t\text{-}1) + v(t) - \hat{\theta}_o(t|t\text{-}1)\{\hat{\theta}_n(t\text{-}1)y(t\text{-}2) + v(t\text{-}1)\}, \qquad t > \tau,$$

since $y(t)$ is defined as:

$$y(t) = \begin{cases} \theta_n(t)y(t\text{-}1) + v(t) & t \le \tau, \\ \theta_n(t)y(t\text{-}1) + v(t) & t > \tau. \end{cases}$$

If sequential substitutions of $y(t)$ continue, then the filter's residual has to be recursively updated. Thus the parameter change detection problem is reduced to the problem of sequential monitoring of residuals to determine when residuals change from white noise to a specific ARMA (Autoregressive Moving Average) process. A suitable F-statistics test can be applied to detect the change point. These techniques are discussed from an estimation standpoint in Broemeling and Tsurumi (1987).

7.3.2 Interaction of estimation and control.

Consider a linear first order scalar system with an additive error $v(t)$:

$$y(t+1) = a_0y(t) + b_0u(t) + v(t+1) \qquad (7.3.12a)$$

where $\theta = (a_0,b_0)'$ is the parameter vector, $u(t)$ a control variable, and $v(t)$ is normally and independently distributed with zero mean and unit variance. For each θ, the variance of the output process

$$J = \lim_{n \to \infty} \left(\frac{1}{N} \sum_{t=0}^{N-1} y^2(t) \right) \tag{7.3.12b}$$

is minimized by the decision rule

$$u(t) = -(a/b)y(t)\,. \tag{7.3.12c}$$

Since θ is unknown, we employ the maximum likelihood method (which is also the least square method here since the error is normally distributed) to obtain the estimate:

$$\hat{\theta}(t) = \left[\sum_{i=0}^{t-1} \phi(i)\phi'(i) \right]^{-1} \sum_{i=0}^{t-1} \phi(i) y(i+1)$$

where $\phi(i) = (y(i), u(i))'$. Then we can use the certainty equivalent control

$$u(t) = -\frac{\hat{a}(t)}{\hat{b}(t)} y(t) \tag{7.3.12d}$$

where $\hat{\theta} = \left(\hat{a}(t), \hat{b}(t)\right)'$. Thus the adaptive control law above can be sequentially computed. In the more general case of an AR(p) model, we have:

$$y(t+1) = \sum_{i=0}^{p} a_i y(t-i) + \sum_{i=0}^{p} b_i u(t-i) + v(t+i), \quad b_0 \neq 0.$$

The estimate $\hat{\theta}$ of the parameter vector $\theta = (a_0, a_1, \ldots, a_p; b_0, b_1, \ldots, b_p)'$ may be obtained by least squares as

$$\hat{\theta}(t+1) = \hat{\theta}(t) + R^{-1}(t)\phi(t)[y(t+1) - \phi'(t)\hat{\theta}(t)]$$

$$R(t) = \sum_{i=0}^{t} \phi(i)\phi'(i)$$

where $\phi(t) = [y(t), y(t\text{-}1), \ldots, y(t\text{-p}); u(t), u(t\text{-}1), \ldots, u(t\text{-p})]'$. The resulting adaptive control law is a bit complicated to derive because of the matrix term $R(t)$. On replacing $R(t)$ by its trace element $r(t)$ we can specify a stochastic recursive algorithm

$$\hat{\theta}(t+1) = \hat{\theta}(t) + \frac{\mu\phi(t)}{r(t)}\left[y(t) - \phi'(t)\hat{\theta}(t)\right] \tag{7.3.12e}$$

where $r(t) = \sum_{i=0}^{t} \phi'(i)\phi(i)$ and μ is a scaling constant. The certainty equivalent control law now can be explicitly computed as

$$u(t) = -\frac{1}{\hat{b}(t)}\left[\sum_{i=0}^{p} a_i(t)y(t-i) + \sum_{i=1}^{p} \hat{b}_i(t)u(t-i)\right]$$

with $\hat{\theta}(t) = \left[\hat{a}_0(t), \hat{a}_1(t), \cdots \hat{a}_p(t); \hat{b}_0(t), \hat{b}_1(t), \cdots, \hat{b}_p(t),\right]'$.

One may note several features of this type of stochastic algorithm. First of all, it has computational simplicity in sequential updating of the parameter estimate and the optimal control. Secondly, the estimate $\hat{\theta}(t)$ can be written in the form of a stochastic gradient algorithm by rewriting (7.3.12e) as $\hat{\theta}(t+1) = \hat{\theta}(t) - \frac{\mu}{2r(t)}\overline{V}_\theta V\left(\hat{\theta}(t)\right)$ where $\overline{V}_\theta V\left(\hat{\theta}(t)\right)$ is a gradient of the loss function $L = \sum_{i=0}^{t}\left[y(t) - \phi'(t-1)\theta\right]^2$ which is minimized by the least square method. The quantity $\mu/2r(t)$ can be regarded as the step size. If $r(t)$ tends to infinity, the step size has a limit of zero and in this case the algorithm converges. Finally, the recursive estimator $\hat{\theta}(t)$ has several optimal statistical properties under the assumption that the error process $v(t)$ is normally and independently distributed. Goodwin and Sin (1984) have applied this statistical theory of optimal design in such cases, where the interaction between the optimal estimates $\hat{\theta}(t)$ and the adaptive control $u(t)$ follows a recursive pattern. Clearly in dynamic economic models many parameters are time varying e.g., the impact of technological innovation in production, such recursive methods of estimation would provide powerful instruments for designing robust control policies.

7.4 Estimating Transversality Conditions

An optimal control system generally leads to a two point boundary condition problem. For a discrete time model, the initial and terminal conditions provide these two sets of boundary points. After the optimal control is applied to the output in the LQG setup, the optimal system is given by a set of difference equations. In the scalar case, this difference equation, also known as the Euler equation, is of the second order and we need two boundary conditions to fully identify the solution. One boundary condition is provided by the initial value of the output, and the other by the terminal value, provided these two values are fixed and known. When the terminal value is free e.g., in the infinite horizon case, fixing its level beforehand may not be feasible. Transversality conditions refer to those conditions by which the free terminal value would lie on the optimal trajectory. These conditions, often stated in terms of the adjoint (also called costate) variables, or the Lagrange multipliers associated with the dynamic equations of motion seek to guarantee three things: Uniqueness of the optimal trajectory, stability in the sense of convergence of the optimal path, and a condition of perfect foresight of the future. In deterministic linear quadratic models of optimal control, these transversality conditions can be imposed fairly easily, but difficulties arise as soon as we enter the field of stochastic control, where the stochasticity may enter as disturbances in the difference equation of motion, or as the standard errors in the parameter estimates of the model. Two most important classes of examples involving the terminal conditions occur in linear dynamic models of economic systems. One is in estimating the Euler equations where the terminal conditions also have to be estimated. The other is in dynamic rational expectations (RE) models, where there are expectational variables which are unobserved.

We consider the case of estimating the Euler equations first and discuss the various approaches of incorporating the transversality conditions as terminal conditions. We start with the approach of Minford et. al. (1979) who, in their Liverpool macroeconomic model, suggested the imposition of terminal conditions only if these characterize the equilibrium properties of the model. For example, they obtained such conditions in their Liverpool model by the analysis of a small linear system on which the complete nonlinear model is based. An alternative method is to set the terminal value by some type of automatic rule. For instance, Fisher (1992) has advocated this procedure and consider three types of impositions. Let $\hat{x}(t)$ denote the solution of the optimal Euler equation in the output space, when a terminal condition is imposed in a scalar model. Then these choices are as follows:

(i) *Fixed value*; $\hat{x}(T+1) = d_{T+1}$, when $\hat{d}_{T+1}$ is a preassigned value. If $\hat{x}(t)$ is already measured as the deviation from some equilibrium value, then d_{T+1} may be set to zero. It is clear that in case of an estimated model, the choice of this terminal value may lead to an error $e(T) = x(T) - \hat{x}(T)$ where $x(T)$ is the equilibrium population value resulting from the dynamic model.

(ii) *Constant level*; $\hat{x}(T+1) = \hat{x}(T)$. This may sometimes run counter to the analytical or equilibrium solution of the optimal Euler equation. Only if the constant level happens to equal the equilibrium solution, that the error $e(T)$ would be zero. Also for nonlinear models this may be more difficult to prescribe.

(iii) *Constant growth*; $\hat{x}(T+1) = \gamma\hat{x}(T)$. Here, γ may be preassigned in terms of the ratio $\hat{x}(T)/\hat{x}(T-1)$. Such preassigned growth rates were advocated at one time in many turnpike models of growth theory by the national planning models. Here also there is the scope of divergence of $\hat{x}(T)$ from its optimal value $x(T)$ and the sensitivity of the optimal trajectory due to such errors may need to be analyzed by simulation experiments.

Recently, Fisher (1992) has considered three large scale econometric models e.g., the LPL (Liverpool) model, the LBS (London Business school) model, and the NIESR (National Institute of Economic and Social Research) model and the implications of specifying the alternative terminal conditions. His experiences with the simulation experiments for the models above reached several broad conclusions. First of all, the imposition of a fixed terminal value works well for stable linear dynamic systems where the characteristic roots are all less than unity in absolute value. However if one is not sure whether the controlled model is stable or not, then sensitivity tests must be used not only to validate the choice of terminal condition in a specific form but also to establish the stability of the model. This recommendation is most important for nonlinear systems for which we cannot obtain the characteristic roots for testing their stability. Secondly, for models with unstable roots near unity, the effectiveness of the terminal condition is considerably improved if the long run solution is known a priory, since these are found to provide better terminal values than those constructed from the solution itself. Finally, the three models exhibit different patterns of sensitivity of the optimal trajectory in respect of the fixed terminal condition. For example for the LPL model, several solutions over a short period are compared with a longer run solution, in which the equilibrium conditions were applied. The terminal conditions applied to the shorter period forecast did not perform well, displaying a substantial divergence for up to 6 years. The terminal conditions based on constant growth or constant levels do much better. The LBS model exhibits a seasonal pattern for the sensitivity of the optimal trajectory, i.e. the trajectory depends on which quarter of the year the simulation considers as the terminal period. The NIESR model exhibits a seasonal pattern of sensitivity as in the LBS model but what is most important is that it reflects the pervasive influence of the exchange rate equation coupled with a unit root. Because the unit root allows any choice of terminal value, no specific form of the terminal condition is attempted. We need sufficient exogenous information to tie down the terminal value.

The rational expectations model postulates a set of expectational variables which contain the agent's expectations of future outcomes. For example, consider the linear dynamic RE model

$$y_t = Ay_{t-1} + Bu_t + Dy_t^e + \varepsilon_t \tag{7.4.1}$$

where y_t^e is a vector of rational expectations of endogenous variables e.g., expectations about the future period $t+1$ formed in the current period and ε_t is the random shock.. Since y_t^e is not observable, one has to hypothesize as to how it can be related to the observed y_t. Two approaches are possible. One is the backward looking view e.g., $y_t^e = \mathrm{E}[y_t | \Omega_{t-1}]$, where Ω_{t-1} is the information set available up to period t-1. In this case, y_t^e can be expressed as a linear combination of lagged values of y_t e.g., $y_t^e = \sum_{i=1}^{p} \alpha_i y_{t-i}$. The second case is the forward looking view e.g., $y_t^e = \mathrm{E}[y_{t+1} | \Omega_t]$. In this case, y_t^e would be a linear combination of future state variables $y_{t+1}, y_{t+2}, \ldots , y_{t+p}$. The most important implication of dynamic RE models arises in the forward looking case. A simple scalar example of such a model is:

$$y_t = ay_{t-1} + bu_t + dy_{t+1} + \varepsilon_t \tag{7.4.2}$$

The effect of the coefficient d in (7.4.2) is very critical in the long run as the terminal time becomes infinitely large. If it makes the system unstable in the sense that at least one characteristic root exceeds one in absolute value, then imposing a fixed terminal value may not be consistent with the optimal trajectory. On the other hand, if the coefficient d preserves the stability of the system, then a fixed terminal condition can be added without much difficulty.

To illustrate some of the basic issues posed by the terminal conditions, we consider now two economic examples, one from the dynamic adjustment model discussed in chapter 5, and the second from the neoclassical theory of optimal growth.

7.4.1 A Dynamic Adjustment Model.

Consider the scalar control model where the objective function already incorporates the control variable:

$$\min J = \mathrm{E}\left[\sum_{t=0}^{\infty} r^t \left\{\alpha\left(y_t - y_t^0\right)^2 + \left(y_t - y_{t-1}\right)^2\right\}\right] \tag{7.4.3a}$$

where y_t is the state variable, E_t is expectation as of time t, $r > 1$ is the exogenous rate of discount and y_t^0 is the desired target level. The first component of the loss function above is the disequilibrium cost due to deviations from the desired target and the second component characterizes the agent's aversion to output fluctuations. The first order conditions for the problem yields the Euler equation:

$$\Delta y_t = r\mathrm{E}_t[\Delta y_{t+1}] - \alpha(y_t - y_t^0) \tag{7.4.3b}$$

where $\Delta y_t = y_t - y_{t-1}$ and $\mathrm{E}_t[.]$ is the conditional expectation with respect to information available up to t i.e., Ω_t. The RE hypothesis involves the expectational variable $E_t[\Delta y_{t+1}]$. Several types of formulations are possible. For example, the backward looking view would replace this by a term involving a linear combination of lagged values of $y_t, y_{t-1}, \ldots, y_{t-p}$ while the forward looking view would consider future values of y_t. But since the future values are not known as of time t, a compromise is needed for all practical situations. A common assumption frequently followed e.g., Kennan (1979) is to consider

$$\Delta y_{t+1} = y_{t+1} - y_t + \zeta_t \tag{7.4.3c}$$

where ζ_t is the error term assumed to be white noise with zero means, in which case $E_t[\Delta y_{t+1}] = y_{t+1} - y_t$. On using this assumption in (7.4.3b), one can solve the second order difference equation:

$$r y_{t+1} - (1 + r + \alpha)y_t + y_{t-1} = -\alpha y_t^0 \tag{7.4.3d}$$

in terms of the characteristic equation of the homogeneous part i.e.

$$r\lambda^2 - (1 + r + \alpha)\lambda + 1 = 0. \tag{7.4.3e}$$

If the desired target level y_t^0 is a constant y^0 and λ_1, λ_2 are the two characteristic roots of (7.4.3e), assumed distinct, the solution can be written very simply as:

$$y_t = A_1\lambda_1^t + A_2\lambda_2^t + y^0 \tag{7.4.3f}$$

where for a finite terminal time T we have:

$$A_1 = (\lambda_2^T - \lambda_1^T)^{-1}[\lambda_2^T(y_0 - y^0) - (y_T - y^0)]$$

$$A_2 = (\lambda_2^T - \lambda_1^T)^{-1}[(y_T - y^0) - \lambda_1^T(y_0 - y^0)]$$

and y_0, y_T are the initial and terminal values of y_t. In case the desired levels y_t^0 are time varying and the time horizon is infinite, we have to make sure that the two roots of (7.4.3e) are stable. If both roots are less than one in absolute value, then the optimal solution defines a stable system and the estimates of the roots from the difference equation (7.4.3d) parameters can be utilized. However in this case the two roots are:

$$\lambda_1, \lambda_2 = (2r)^{-1}[(1+r+\alpha) \pm \{(1+r+\alpha)^2 - 4r\}]$$

where the product of the roots is r^{-1}. Since $r > 1$, the roots are positive, with one root greater than one, and the other smaller than one. Let λ_1 be the stable root (i.e. $\lambda_1 < 1$). The unstable root λ_2 makes the optimal trajectory explosive (i.e. unbounded) as t becomes infinite. Hence one has to choose the stable root in order to assure stability of the optimal path. Thus on using only the stable root, the second order Euler equation is transformed a first order equation:

$$y_t = \lambda_1 y_{t-1} + d, \tag{7.4.3g}$$

where

$$d = (1 - \lambda_1)(1 - r\lambda_1)E_t\left[\sum_{t=0}^{\infty}(r\lambda_1)^t y_t^0\right]. \tag{7.4.3h}$$

If d is a bounded constant $\bar{d}$ in the limit as t becomes infinite, then y converges to the steady state solution:

$$y = \lim_{t\to\infty}\left[\left(y_0 - \frac{\bar{d}}{1-\lambda_1}\right)\lambda_1^t + \frac{\bar{d}}{1-\lambda_1}\right] = \frac{\bar{d}}{1-\lambda_1}.$$

Two general methods of estimation are possible here due to the expectational variable $E_t[.]$. One is to replace Δy_{t+1} by the stochastic term $(y_{t+1} - y_t + \zeta_{t+1})$ as in (7.4.3c) where $E_t\eta_t = 0$ for all t, and rewrite (7.4.3b) as:

$$\Delta y_t = r\Delta y_{t+1} - \alpha(y_t - bz_t) + v_t \tag{7.4.3i}$$

where the error term is $v_t = (r\eta_{t+1} + \alpha e_t)$ and it is assumed that $y_t^0 = bz_t + e_t$. Here the stochastic target variable y_t^0 is assumed for simplicity to be linearly related to an observed exogenous variable z_t and e_t is assumed to be a white noise error component, independent of η_t thus implying $E_{t-1}[v_t] = 0$. Hence one could apply the method of instrumental variables to estimate the coefficients of (7.4.3g) consistently i.e. by ordinary least squares when the errors are Gaussian. Kennan has used several exogenous variables including time in place of only one i.e. z_t in (7.4.3i).

A second method of estimation employs an assumption about how z_t is generated and thus derives an estimating equation. For example assume that z_t follows a first order process: $z_t - z_{t-1} = \phi_t$, where ϕ_t is a white noise error term such that $E_{t-1}[\phi_t] = 0$. This yields the estimating equation

$$y_t = \lambda_1 y_{t-1} + (1 - \lambda_1)bz_t + (1 - r\lambda_1)(1 - \lambda_1)e_t.$$

In case z_t follows a second order process (i.e. $z_t - 2z_{t-1} + z_{t-2} = \phi_t$), the estimating equation becomes $\Delta y_t = (\lambda_1 - 1)(y_{t-1} - bz_t) + (1 - r\lambda_1)(1 - \lambda_1)b\Delta z_t + (1 - r\lambda_1)(1 - \lambda_1)e_t$, and so on for the third order process.

While these two methods of estimation can be employed with any given data set, the simulation studies by Pagan and Smith (1993) found the first method to be more robust. Note that this model can easily be generalized to the case when y_t and y_t^0 are vectors. Imposing transversality conditions implies rejecting all the unstable roots and therefore reduce the order of the system.

7.4.2 A Model of Optimal Growth.

We now consider a neoclassical model of optimal growth in a finite horizon case;

$$\max_{c_t} J = \sum_{t=0}^{T} r^{-t} U(c_t) \qquad \text{subject to:} \quad c_t = f(k_t) + k_t - (1 - n)k_{t+1} \quad (7.4.5)$$

where the control variable c_t is consumption per capita, k_t is capital stock per capita, and the growth of labor $n = (L_{t+1} - L_t)/L_t$ is constant. The production function $f(k_t)$ is non linear and concave satisfying the neoclassical assumption of constant returns to scale. The two boundary conditions are $k(0) = k_0$, $k(T) = k_T$. On assuming the logarithm form for the concave utility function $U(c_t) = \ln\,(c_t - c_0)$ where c_0 is a given level of initial consumption, the Euler equation for the optimal trajectory can be easily derived as

$$(c_t - c_0)^{-1}[1 + f_k(k_t)] - (1 - n)r(c_t - c_0)^{-1} = 0 \qquad (7.4.6)$$

where $f_k(k_t) = \dfrac{\partial f(k_t)}{\partial k_t}$ is the marginal productivity of capital. The steady state solution or the turnpike exists when we have solutions for the non linear steady state equation:

$$1 + f_k(\bar{k}) = (1 + n)r \qquad (7.4.7)$$

where $\bar{k}$ is the steady state value of k. Note that if this steady state equation has more than one solution, then even if the dynamic system (7.4.6) is convergent, it may not converge to a unique steady state since there are multiple steady states.

Two types of approaches are possible in such a situation. One is to accept the proposal of Minford and Peel (1983) to adopt the criterion of the "most stable path" (i.e. the roots with the smallest modulus for a linearized version of the nonlinear path) and which is free of extraneous variables. However, for nonlinear optimal trajectories there may be difficulties when there may be more than one way to linearize. A second approach is to prescribe the terminal conditions exogenously. A simulation approach which evaluates the sensitivity of the optimal trajectory is also adopted by many practitioners. Minford and Peel have discussed some closely related methods other than those discussed here e.g., the method of undetermined coefficients by Muth and Lucas, Sargent's method of forward substitution, and the method of equilibrium solution. Recently, Hall and Henry (1988) examined the problems of specifying the terminal conditions empirically in terms of the nonlinear econometric model prepared by the UK National Institute of Economic and Social Research over quarterly data for the period 1983-87. They studied the imposition of the following terminal conditions:

(i) Setting the terminal conditions to exogenously determined levels, i.e. the exogeneity rule.

(ii) Setting the terminal conditions so that they lie on the previous year's growth path, i.e. the output level rule

(iii) Setting the terminal conditions so that the levels of the state or output variables are projected flat, i.e. the sensitivity rule, and

(iv) Setting the terminal conditions so as to project a constant rate of growth from the final quarter of the solution period, i.e. constant growth rule.

Their empirical experience is that the early parts of the computation run are not substantially affected by the form of the terminal conditions as above. However the picture becomes altered radically when a forward looking exchange rate equation is added to the model. In this case, it proved almost impossible to solve the model except with fixed terminal conditions i.e. the exogeneity rule above.

We may propose a somewhat new approach to estimate the terminal condition in the simple case where the Euler equation is linear and with constant coefficients. For example consider the growth model (7.4.5) with a linear production function: $f(k_t) = \beta k_t$ and assume for simplicity that c_t is measured as a deviation from c_0. then the homogeneous part of the Euler equation would appear as:

$$(1+n)k_{t+1} + \left(\frac{(1+\beta)(1+r)}{r}\right)k_t - \left(\frac{(1+\beta)^2}{r(1+n)}\right)k_{t-1} = 0 \tag{7.4.8}$$

The two roots are:

$$\lambda_1, \lambda_2 = \left(\frac{(1+\beta)(1+r)}{r} \pm \frac{(1+\beta)\sqrt{(1+r)^2+4r}}{r}\right)\left(\frac{1}{2(1+n)}\right) \tag{7.4.9}$$

for a finite terminal time $T+1$ the optimal solution is

$$k_t = A_1\lambda_1^t + A_2\lambda_2^t \tag{7.4.10}$$

where $A_1 = (\lambda_2^{T+1} - \lambda_1^{T+1})(k_0\lambda_2^{T+1} - k_{T+1})$, and $A_1 = (\lambda_2^{T+1} - \lambda_1^{T+1})(k_{T+1} - k_0\lambda_1^{T+1})$. Assuming a known value for the discount rate r and for n, we estimate the parameter $\hat{\beta}$ from (7.4.8), and using this estimate compute the two characteristic roots λ_1, λ_2. Finally we estimate the two parameters k_0 and k_{T+1} from (7.4.10). If

k_0 is known, then one could apply ordinary least squares to estimate the parameter k_{T+1} from

$$k_t = \gamma_{0t} + k_{T+1}\gamma_{1t}$$

where γ_0 and γ_t are known or estimated time series $\gamma_{0t} = (\lambda_2^{T+1} - \lambda_1^{T+1})^{-1}k_0(\lambda_1^t - \lambda_2^t)$, and $\gamma_{1t} = (\lambda_2^{T+1} - \lambda_1^{T+1})^{-1}k_0(\lambda_1^t - \lambda_2^t)$. For the infinite horizon case however we have to reject the unstable root and thereby reduce the second order Euler equation to first order.

Clearly the problem of efficient estimation of the transversality condition or its impact on the Euler equation provide an open field of research for the applied econometrician.

7.5 Concluding remarks

Applied control theory has now entered into a new phase. An increasing number of nonlinear control models are being applied in modeling economic dynamics. Chaotic dynamics and complexity theory, entropy and information theory and even distributed parameter systems where partial differential equation systems are used instead of the ordinary differential equation used to model the economic dynamics. Invariably these new concepts would require developing efficient methods for econometric estimation, particularly for nonlinear systems. Simulation and experimental analysis of structural data appear to be very useful in this framework.

References

Beedles, W. (1979), "On the Asymmetry of Market Returns." *Journal of Financial and Quantitative Analysis*, vol 14, pages 653-660.

Bera, A., Higgin, M. and Lee, S. (1992), Interaction Between Autocorrelation and Conditional Heteroscedasticity: A Random Coefficient Approach." *Journal of Business and Economic Statistics*, vol. 10, pages 133-142.

Black, F. (1976), "Studies fo Stock Market Volatility Changes." *Proceedings of the American Statistical Associaton*, Vol. 71, pages 177-181.

Broemeling, L. and Tsurumi, H. (1987), *Economerics and Structural Change*, New York: Marcel Decker.

Chow, G. (1975), *Analysis and Control of Dynamic Systems*, New York: John Wiley.

Fisher, P., (1992) *Rational Expectations in Economic Models*, Kluwer, Advanced Series in Theoretical and Applied Econometrics, Dortrecht, The Netherlands.

Goodwin, G. and Sin, K., (1984), *Adaptive Filtering, Prediction and Control.* Englewood Cliffs, Prentice Hall.

Green, C. (1988), "Adjustment Costs and Mean Variance Eficiency in U.K. Financial Markets." *Economic modelling in OECD countries*, London: Chapman and Hall.

Gregory, A., Pagan, A. and Smith, G., (1993), "Estimating Linear Quadratic Models with Integrated Processes." P. Phillips ed., *Models, Methods and Applications in Econometrics*, Cambridge: Blackwell Publishers

Hall, S. And Henry, S., (1988), *Macroeconomic Modelling*, Amsterdam: North Holland.

Karlin, S. (1982), "Some Results on te Optimal Partitioning of Variance and Monoticity With Truncation Level." *Statistics and Probability*, Amsterdam: North Holland.

Kennan, J. (1979). "The Method of Partial Adjustment Models With Rational Expectations." *Econometrica*, vol. 47, pages 1441-1457.

Ljung, L., (1987). *System Identificarion, Theory for the User*. Englewood Cliffs, Prentice Hall.

Ljung, L., and Soderstrom, T. (1974) "Identification of Linear Multivariable Systems Operating Under Linear Feedback Control." *IEEE Transactions on Automatic Control*, AC-19, pages 836-840

Minford, P. and Peel, D. (1983), *Rational Expectations and the New Macroeconomics*. London: Basil Blackwell.

Minford, A., Matthews, K. and Marwaha, S. (1979), "Terminal Conditions, Uniqueness, and the Solutions for Rational Expectations Models with Forward Expectations." *Economic Letters*, vol. 4, pages 117-120

Nelson, D. (1991) "Conditional Heteroscedasticity: a New Approach."*Econometrica*, vol. 59, pages 347-70.

Pindyck, R. (1988), "Capital Risk and Models of Investment Behavior." *Economic Modelling in OECD Countries*, London: Chapman and Hall.

Richmond, J. (1974), "Estimating the Efficiency of Production." *International Economic Review*, vol. 15, pages 515-521.

Saunders, A., Ward, C. and Woodward, R. (1980), "Stochastic Dominance and the Performance of U.K. Unit Trusts." *Journal of Financial and Quantitative Analysis*, vol. 15, pages 323-330.

Sengupta, J. and Sfeir, R., (1979), "The Adjustment of Output Inventory Process Under Linear Decision Rules." *Journal of Economic Dynamics and Control*. vol.1, pages 361-381.

Whittle, P. (1990), *Risk Sensitive Optimal Control*. New York: John Wiley.

APPENDIX: Some useful results in matrix algebra

Eigenvalues and eigenvectors.

Theorem: (Cayley-Hamilton).
A square matrix is solution of its own characteristic polynomial.

Theorem: (Gershgorin)
The eigenvalues of a square complex matrix Z of dimension n by n lie within the union of the n circles in the complex plane, with centers z_{ii} and with radius $\sum_{\substack{j=1 \\ j\neq i}}^{n} \left| z_{ij} \right|$, i = 1,... n.

Note: The theorem is also valid if $|z_{ij}|$ is replaced by $|z_{ji}|$, i.e. if the summation is done over the columns of Z instead of its rows.

Function of matrices.
There are two approaches to define a function of a matrix; the power series approach, and the spectral approach. The power series approach is simply an adaptation of power series for a scalar and is the most commonly used. The series are defined only if they are convergent, which can be difficult to establish without the use of a computer. Recalling that $(1-a)^{-1} = \sum_{i=0}^{\infty} a^i$, we can define the matrix function:

$$(1-A)^{-1} = \sum_{i=0}^{\infty} A^i .$$

Similarly, since $e^a = I + a + \frac{a^2}{2!} + \frac{a^3}{3!} + \cdots = \sum_{i=1}^{\infty} \left(\frac{a^i}{i!} \right)$, we can also define:

$$e^A = I + A + \frac{A^2}{2!} + \frac{A^3}{3!} + \cdots = \sum_{i=1}^{\infty} \left(\frac{A^i}{i!} \right).$$

The power series approach is somewhat cumbersome because it replaces a simple expression by an infinite series. The spectral approach is a bit more involved, but yields very elegant results that can be used easily.

Definition: The *spectrum* of a square matrix is the set of all eigenvalues of the matrix.

Theorem: Let $\Omega = \{\lambda_1, \lambda_2, \cdots \lambda_k\}$ be the set of distinct eigenvalues of a square matrix A, with respective multiplicity m_i . Let $f(\lambda)$ be a function defined on Ω. If there exists a polynomial $g(\lambda)$ such that $g^{(j)}(\lambda_i) = f^{(j)}(\lambda_i)$ for $i = 1 \cdots k$ and $j=0 \cdots (m_i-1)$, then the matrix-valued function $f(A)$ is defined as $f(A) = g(A)$.

The proof of this theorem, which relies on the Caley Hamilton theorem, is beyond the scope of this book. For more details and examples, the reader could consult C.T. Chen (1970). The theorem is quite powerful as it allows us to express many functions as a polynomial of a matrix which usually can be computed easily. Note that the derivatives are taken with respect to the eigenvalues, and not with respect to the variables in the matrix A.

Example 1:

Given $A = \begin{bmatrix} 0 & 0 & -2 \\ 0 & 2 & 0 \\ 1 & 0 & 3 \end{bmatrix}$. Compute: $f(A) = e^{At}$.

The eigenvalues of A are the roots of the characteristic polynomial:
$\det(A - \lambda I) = -\lambda(2 - \lambda)(3 - \lambda) - 2(2 - \lambda) = (2 - \lambda)^2(1 - \lambda)$.
The roots are $\lambda_1 = 2$, with multiplicity 2, and $\lambda_2 = 1$, with multiplicity 1.
We can now use the theorem above, with $\Omega = \{2, 1\}$, $m_1 = 2$, $m_2 = 1$. First, we select the polynomial $g(\lambda) = \alpha_0 + \alpha_1 \lambda + \alpha_2 \lambda^2$ which must be of order 3 since we have three conditions to meet:

$$\begin{aligned} f(\lambda_1) &= g(2); & e^{2t} &= \alpha_0 + 2\alpha_1 + 4\alpha_2 \\ f'(\lambda_1) &= g'(2); & e^{2t} &= \alpha_1 + 4\alpha_2 \\ f(\lambda_2) &= g(1); & e^{t} &= \alpha_0 + \alpha_1 + \alpha_2 \end{aligned}$$

We then solve the system above to identify the coefficients of $g(\lambda)$; The first two equations yield: $\alpha_0 + \alpha_1 = 0$, hence the third one yields: $\alpha_2 = e^t$, and $\alpha_0 = -\alpha_1 = 4e^t - e^{2t}$. Finally, we compute $f(A) = g(A) = \alpha_0 I + \alpha_1 A + \alpha_2 A^2$;

$$f(A) = (4e^t - e^{2t}) I - (4e^t - e^{2t}) A - 4e^t A^2$$

After substitution of I, A, and A^2, we have;

$$f(A) = e^{At} = \begin{bmatrix} 12e^t - e^{2t} & 0 & 32e^t - 2e^{2t} \\ 0 & -4e^t + e^{2t} & 0 \\ -16e^t + e^{2t} & 0 & -32e^t + 2e^{2t} \end{bmatrix}.$$

Norm of vectors and matrices.

A norm is a unique number attached to a vector or a matrix, which relates the size of their elements. Norms are not unique and the concept of length can be viewed as one of the many possible norms. The norm of a vector x (or matrix) defined on the set of complex numbers is denoted by $\| x \|$. Norms must satisfy three relationships;

1. $0 \le \| x \|$, and $\| x \| = 0$ if and only if $x = 0$.
2. $\| \alpha x \| = | \alpha | \, \| x \|$.
3. $\| x_1 + x_2 \| \le \| x_1 \| + \| x_2 \|$.

Three vector norms are commonly used;

- The *Absolute Value norm*: $\| x \|_1 = \sum_{i=1}^{n} |x_i|$
- The *Euclidian norm*: $\| x \|_2 = \sqrt{\left(\sum_{i=1}^{n} |x_i|^2 \right)}$
- The *Infinity norm*: $\| x \|_\infty = \max | x_i |, \quad i = 1, 2, \cdots n$.

The subscripts 1, 2, and ∞ are commonly used to refer to the Absolute Value, Euclidian and Infinity norms (i.e. the *2-norm* is the Euclidian norm). The norm of a matrix (A) is defined by the ratio of the norm of the vector Ax divided by the norm of the vector x;

$$\| A \| = \sup_{x \neq 0} \frac{\|Ax\|}{\|x\|} = \sup_{\|x\|=1} \|Ax\|,$$

where the supremum is taken over any non zero vector x in the domain of definition of A. It is also convenient to restrict the possible vectors x to have unit norm, in which case the norm of a matrix can be defined solely by its components;

- Absolute Value norm: $\| A \|_1 = \max_j \left(\sum_{i=1}^{n} |a_{ij}| \right)$

- Euclidian norm: $\| A \|_2 = \max_j \sqrt{\left(\sum_{i=1}^{n} \left|a_{ij}\right|^2\right)}$

- Infinity norm: $\| A \|_\infty = \max_i \left(\sum_{j=1}^{n} \left|a_{ij}\right|\right)$

The norm of a matrix has the properties:

1. $\| A + B \| \leq \| A \| + \| B \|$
2. $\| A B \| \leq \| A \| \, \| B \|$

Matrix inversion lemma:

For any non singular matrix A partitioned as: $A = \begin{bmatrix} A_{11} & A_{12} \\ A_{21} & A_{22} \end{bmatrix}$, the following relationship holds; $\left(A_{11}^{-1} + A_{12}A_{22}A_{21}\right)^{-1} = A_{11} - A_{11}A_{12}\left(A_{21}A_{11}A_{12} + A_{11}^{-1}\right)^{-1} A_{21}A_{11}$

Systems of differential equations.

Theorem[19]:

Let $A(t)$ be a continuous matrix-valued function defined on $\Re^n$ such that $A(t_1)A(t_2) = A(t_2)A(t_1)$ for all real numbers t_1, t_2 . Let $c = (c_1, c_2, \cdots c_n)$ be a vector of complex numbers. Then the system $dx/dt = A(t)x(t)$ has solution: - $x(t) = \left(e^{\int_{t_0}^{t} A(\tau)d\tau} \right) x(t_0)$.

19 For a proof, see Mostow, Sampson & Meyer, (1963), p373.

INDEX

—A—

—B—

—C—

—D—

—J—

—K—

—L—

—M—

—R—

—S—

GPSR Compliance

The European Union's (EU) General Product Safety Regulation (GPSR) is a set of rules that requires consumer products to be safe and our obligations to ensure this.

If you have any concerns about our products, you can contact us on ProductSafety@springernature.com

In case Publisher is established outside the EU, the EU authorized representative is:

Springer Nature Customer Service Center GmbH
Europaplatz 3
69115 Heidelberg, Germany

Batch number: 10755292

Printed by Printforce, the Netherlands